THE SUPER BOOK OF
Useless
Information

DI018046

THE SUPER BOOK OF
USELESS
INFORMATION

The Most Powerfully Unnecessary Things
You Never Need to Know

DON VOORHEES

A PERIGEE BOOK

A PERIGEE BOOK
Published by the Penguin Group
Penguin Group (USA) Inc.
375 Hudson Street, New York, New York 10014, USA
Penguin Group (Canada), 90 Eglinton Avenue East, Suite 700, Toronto, Ontario M4P 2Y3, Canada
(a division of Pearson Penguin Canada Inc.)
Penguin Books Ltd., 80 Strand, London WC2R 0RL, England
Penguin Group Ireland, 25 St. Stephen's Green, Dublin 2, Ireland (a division of Penguin Books Ltd.)
Penguin Group (Australia), 250 Camberwell Road, Camberwell, Victoria 3124, Australia
(a division of Pearson Australia Group Pty. Ltd.)
Penguin Books India Pvt. Ltd., 11 Community Centre, Panchsheel Park, New Delhi—110 017, India
Penguin Group (NZ), 67 Apollo Drive, Rosedale, Auckland 0632, New Zealand
(a division of Pearson New Zealand Ltd.)
Penguin Books (South Africa) (Pty.) Ltd., 24 Sturdee Avenue, Rosebank, Johannesburg 2196, South Africa

Penguin Books Ltd., Registered Offices: 80 Strand, London WC2R 0RL, England

While the author has made every effort to provide accurate telephone numbers and Internet addresses at the time of publication, neither the publisher nor the author assumes any responsibility for errors or for changes that occur after publication. Further, the publisher does not have any control over and does not assume any responsibility for author or third-party websites or their content.

First edition: October 2011

Library of Congress Cataloging-in-Publication Data

Voorhees, Don.
  The super book of useless information : the most powerfully unnecessary things you never need to know / Don Voorhees.— 1st ed.
    p. cm.
  ISBN 978-0-399-53696-0
  1. Questions and answers.   I. Title.
  AG195.V65 2011
  031.02—dc23   2011025858

PRINTED IN THE UNITED STATES OF AMERICA

10   9   8   7   6   5

Most Perigee books are available at special quantity discounts for bulk purchases for sales promotions, premiums, fund-raising, or educational use. Special books, or book excerpts, can also be created to fit specific needs. For details, write: Special Markets, Penguin Group (USA) Inc., 375 Hudson Street, New York, New York 10014.

CONTENTS

SUPER SQUIRRELS

Rocky the Squirrel's full name is Rocket J. Squirrel.

Rocky lives with Bullwinkle the Moose in Frostbite Falls, Minnesota.

Rocky's character was voiced by actress June Foray.

Secret Squirrel, also known as Agent 000, was voiced by Mel Blanc.

Secret's sidekick was Morocco Mole.

Secret's archenemy was Yellow Pinkie. The name is a play on the 007 nemesis Goldfinger.

IT'S ALIVE!

OBSESSION FOR CATS

Zookeepers have found that tigers and other big cats are fascinated with the smell of Obsession for Men by Calvin Klein. They spray it on logs to keep the animals amused. The felines are also quite taken with the aroma of rhino urine, which is much cheaper.

Bobcat, mountain lion, coyote, and fox pee are sold commercially to repel deer, mice, and rabbits.

REPTILE ISLE

The Galapagos Islands get their name from the Spanish word for "saddle," referring to the saddle-shaped shell of the Galapagos tortoises that reside there.

The Galapagos tortoise can go for up to a year without eating or drinking, living off its stored fat.

The marine iguanas of the Galapagos Islands are the only lizards that frequent the ocean.

Marine iguanas sneeze to expel the salt they ingest while feeding under the sea.

MOTHER'S SEAL OF APPROVAL

The hooded seal has the shortest lactation period of any mammal. Pups are weaned from the mother's 60 percent fat content milk in a mere four days, after doubling in size.

Female elephant seals will lose six hundred pounds in less than a month, once they begin nursing their young.

POCKET PETS

The pygmy marmoset, or dwarf monkey, that dwells in the jungles of South America is just fifteen centimeters long (not counting the tail).

The dwarf gecko of St. Martin is the world's smallest reptile at three-quarters of an inch in length.

The Brazilian gold frog is about half the size of a dime.

MUSHERS

Inuit sled dogs can travel one thousand miles in eight days. This is the farthest any land animal can travel in so short a time.

These dogs can live on seal blubber and snow alone.

The Inuits used the fur from their Samoyed dogs to spin wool for clothing, as some modern people still do today.

DOGGONE SMART

A six-year-old border collie named Chaser has a "vocabulary" of 1,022 words. The precocious pooch has been trained to fetch over one thousand different items on command and also understands several verbs.

DOGGY DOWNERS

Macadamia nuts, caffeine, grapes, raisins, avocadoes, and sesame seeds are all toxic to dogs.

Dogs with short snouts are more likely to die while traveling on a plane than their long-nosed brethren.

Because they are so big, Irish wolfhounds only live to about six or eight years of age.

In rare cases, people who sleep with their dogs and cats can contract some rather nasty diseases, including bubonic plague (i.e., the Black Death), Chagas' disease, and cat scratch disease. Kissing Fido or Fluffy can likewise spread these diseases.

DUSTUPS

Male bison roll around on dry ground to make as much dust as possible to intimidate other males and attract females.

BEAR NECESSITIES

Some grizzly bears will spend their entire summer eating only moths that they find congregating under rocks high in the Rockies. A bear can consume up to forty thousand moths in one day. That's about twenty thousand calories' worth.

SPEEDY SPECIES

North American cheetahs roamed in what is now Yellowstone National Park as recently as ten thousand years ago.

The North American pronghorn can run at speeds up to sixty miles per hour and is the fastest herbivore on the planet.

COOL CUSTOMERS

Ice worms live on the surface of the glaciers of North American west coast glaciers. These relatives of the earthworm subsist on algae and bacteria.

The metabolism of the worms increases at lower temperatures and they die at temperatures above 50°F.

HERBAL ESSENCE

Feeding cows an extract of oregano reduces their output of methane gas by 40 percent.

LITTLE BUGGERS

Jiggers are flea-like insects that burrow under the skin of humans, where they suck the blood and breed. They commonly attack the buttocks, lips, and eyelids, causing the flesh to rot away.

Eighty percent of the world's insects live in jungles.

GOING APE

Chimpanzee groups go to war with each other and engage in cannibalism of their enemies.

The snub-nosed monkeys of Myanmar have noses so upturned that rain falls into their nostrils and the monkeys must put their heads down between their legs to keep the water out.

BUG BITES

Male bedbugs stab the females through the abdomen to inseminate them in a process known as traumatic insemination. The females tend to flee and hide from the males for this reason, making them hard for exterminators to find.

The average cost to exterminate bedbugs from a two-bedroom apartment is one thousand dollars.

HOW TWEET IT IS

There are more bird species in Costa Rica than in the United States and Canada combined.

The honeyguide is a parasitic bird that lays its eggs in the nests of other birds. When the young hatch, the honeyguide chicks will kill off the host bird's young and destroy the other eggs.

Turkeys can lay eggs that have two yolks.

Pelicans are known to eat the chicks of other birds, particularly gannets.

Clark's grebes, a type of water bird, can literally run on water.

SOMETHING TO CROW ABOUT

Crows use about two hundred and fifty different calls.

Crows use different warning calls for cats, hawks, and humans.

In Japan, crows collect metal coat hangers with which to build their nests.

Young crows stay with their parents for up to five years, the longest of any bird species.

BYE-BYE BIRDIE

During the early nineteenth century, enormous flocks of passenger pigeons darkened the skies of North America east of the Rockies. Some flocks numbered in the billions and stretched one mile in width by three hundred miles in length.

Passenger pigeons were hunted to extinction to feed slaves and the poor.

The last passenger pigeon, named Martha, died in 1914 in the Cincinnati Zoo.

FISH STORIES

The eel-like oarfish, which can grow to lengths of fifty-six feet, are the largest bony fish.

An octopus has nine brains—one in each arm and one in the "head."

FLIPPER FACTS

Dolphins have no vocal cords, but make vocalizations using nasal air sacs located below their blowholes.

Dolphins make whistles and burst-pulsed sounds. They use clicking noises for echolocation.

Dolphins hear through their lower jaws, which transmit sound to the middle ear.

Dolphin teeth are thought to act as antennae helping to pinpoint the source of sounds.

Dolphins have no sense of smell.

Seventy-five percent of wild dolphins bear scars from shark bites.

MUDDY MEALS

Forest elephants ingest special clay mud for needed nutrients and to neutralize the toxins found in the tree leaves that they eat.

WORKING 9 TO 5

"Crepuscular" animals are those that are active from dawn to dusk.

BALLSY BLOOMS

The word "orchid" is from the Greek *órkhis*, meaning "testicle," which the roots somewhat resemble.

Orchids are the second largest family of plants. There are twice as many orchid species as bird species.

Horticulturalists have developed more than one hundred thousand types of orchids.

All wild species of orchids are endangered.

In greenhouses, some orchids can live over one hundred years.

The smallest orchid blooms are microscopic.

Orchid seeds lack an endosperm and must land near a certain fungus that they enter into a symbiotic relationship with in order to get the nutrients required to germinate.

Orchid flowers bloom for a very long time because their pollination systems are so specialized that their chance of being pollinated is low. Thus they stay in flower for as long as possible to increase the odds of attracting the particular pollinator specific to the species.

Some orchids mimic the appearance and smell of female insects, which encourages the male insects to "mate" with the flowers and thus pollinate them.

AMBER WAVES

There are thirty thousand varieties of wheat.

Winter wheat is planted in the fall. The seeds sprout and then go dormant over the winter, until the soil warms in the spring.

DUNG ON A STICK

Mistletoe is a parasitic plant that only grows in the tops of trees.

The name "mistletoe" is from the German for "dung on a twig." This is because the plant's seeds are spread by the droppings of the birds that eat the fruits that contain them.

Mistletoe fruit is poisonous to humans.

TIMBRRR

Subarctic taiga forests are the largest on Earth, having more trees than all the rain forests combined.

Because of the cold climate, the growing season in the taiga may be as short as one month a year and some trees may not grow past the seedling stage for sixty years.

SCIENCE DIGEST

FRUITFUL DISCOVERY

In the Middle Ages, bread mold (*Penicillium*) was used to treat infected wounds.

> Many earlier researchers had noticed the antibiotic properties of the *Penicillium* fungus, long before Alexander Fleming made his 1928 discovery, but they failed to act on it.

Penicillin was originally mass-produced from a fungus found growing on a rotting cantaloupe in a Peoria, Illinois, market.

> It wasn't until 1943 that penicillin could be mass-produced, just in time for use by the Allied troops on D-day.

MAKING WAVES

"Tsunami" is Japanese for "harbor wave." They are sometimes referred to as "tidal waves," but they have nothing to do with the tides.

Japan gets more tsunamis than any other country.

The tsunami that struck Japan in 2011 generated waves of 124 feet that traveled up to 6 miles inland, leaving 28,000 people dead or missing.

Tsunamis most frequently occur in the months of March, August, and November.

In 1946, all the water drained from Hilo Harbor in Hawaii, followed by a tsunami that wiped out the waterfront and killed 150 people.

Nuclear blasts and meteorites can cause tsunamis.

Tsunami waves travel at more than five hundred miles per hour.

Eighty percent of tsunamis occur in the Pacific Ocean.

There was a tsunami that hit Scotland in 6100 BC and one that struck Cornwall, England, in 1755.

Although not tsunamis, there can be thirty-foot waves on the Great Lakes.

RUNNING HOT AND COLD

The average global temperature went up in 2002, down in 2004, up in 2005, down in 2006, up in 2007, down in 2008, and up in 2009. Overall between 2002 and 2009, the average world temperature decreased slightly.

December 2010 was the second coldest December since 1659.

The average mean surface temperature on Earth in 2010 was 59°F.

COOL KIDS

Students at Falls Elementary School in International Falls, Minnesota, have to go outside for recess unless the temperature is below −15°F. Kids at Asheville, North Carolina, schools aren't sent out for recess unless the temperature is *above* 40°F.

WEATHER YOU LIKE IT OR NOT

St. Elmo's Fire is a bright, loud electrical discharge from pointy objects, such as ship's masts, steeples, or airplanes, that occurs during stormy weather.

The highest temperature ever recorded in Canada was 113°F in Yellow Grass, Saskatchewan, in 1937.

Accurate weather data for the United States only goes back to 1880.

Niphablepsia is the fancy name for "snow blindness."

Nephelococcygia is the scientific term for seeing shapes in cloud formations.

Nor'easters can produce winds in excess of hurricane force.

There have been as many as four hurricanes at the same time in the Atlantic Ocean.

A "nowcast" is a very short-term weather forecast for the next few hours.

The National Weather Service does not specify how much of the sky must be covered to qualify as "partly cloudy."

Trace precipitation is an amount less than 0.005 inches.

A typhoon is a hurricane in the western North Pacific Ocean. It is known as a cyclone in the Indian Ocean.

Snowflakes are six-sided because of the hexagonal shape formed when water molecules join together.

The average snowflake is made up of 180 billion molecules of water.

Yellow snow can result from pine or cypress pollen in it, among other reasons.

The mercury in a thermometer will freeze at –38°F. Alcohol thermometers freeze at –175°F.

Dry air is heavier than humid air.

Indian summer is a warm spell in mid- to late autumn, after the first frost.

Fog is when water droplets in the air reduce the visibility to ⅝ mile. Fog occurs when the temperature and dew point are the same (or nearly the same).

Drizzle is slowly falling water droplets that are less than 0.02 inches in diameter.

Mist is composed of microscopic water droplets, which do not lessen visibility as much as fog.

⬛ SLIPPERY WHEN WET

Weather is a factor in 24 percent of car crashes.

The odds of having a car accident are 70 percent higher in the rain.

SPOUTING OFF

Florida Bay is considered the "waterspout capital" of the world. Nearly five waterspouts a year occur there. They also occur on the Great Lakes.

Most waterspouts are not as strong as tornadoes.

Waterspouts do not suck up water. The water seen in the spout comes from condensation.

Waterspouts begin as a dark spot on the surface of the water, followed by a whirlpool-like, spiral pattern on the water's surface, formation of a spray ring, and finally development of the spout.

DOWN THE DRAIN

The two most powerful maelstroms, or whirlpools, are found in Norway. The third most powerful maelstrom is Old Sow, located between Eastport, Maine, and New Brunswick, Canada.

Some maelstroms have strong downdrafts capable of pulling a person wearing a life jacket down to depths of one thousand feet before spitting them back up far down current.

SANDY STORY

The word "sahara" comes from the Arabic *sahrā*, meaning "desert." Calling it the Sahara Desert is thus redundant.

In the eastern Sahara, 97 percent of the days are sunny.

There are sand dunes in the Sahara that reach heights of almost six hundred feet.

The dunes of the Sahara make sounds similar to that of blowing the top of an open soda bottle. They are the result of the friction of the sand particles moving within the dunes and can be heard for up to ten kilometers away. The Arabs used to believe the noises were made by invisible creatures.

Thousands of years ago, the Sahara was a much wetter place, supporting hippopotamuses.

One-third of the earth's land is covered by desert.

THE ASCENT OF MAN

In the last twenty thousand years the size of the human brain has shrunk by 10 percent.

Around 7,500 years ago, Europeans began to evolve a tolerance for lactose.

Sickle-cell anemia evolved within the past three thousand to four thousand years to protect humans from malaria. The disorder deforms red blood cells into a sickle shape, making it hard for the malaria parasite to infect them. Eight percent of African Americans have the trait for sickle-cell anemia, as compared to one in two thousand to one in ten thousand Caucasians.

The cerebral cortex of the human brain is remarkably like a clump of neurons inside the head of a marine ragworm. The similarities are so close that they cannot be explained by any reason other than humans and the worms share a common ancestor.

KNUCKLE DRAGGERS

The term "Cro-Magnon" comes from the rock shelter in France of the same name, where the first specimen of early modern humans was discovered.

The last Neanderthal died about thirty thousand years ago.

Scientists have recently found Neanderthal DNA in a small percent of modern humans. This means humans and Neanderthals interbred about sixty thousand years ago.

> The famous Australopithecus skeleton, Lucy, was named for the Beatles song "Lucy in the Sky with Diamonds," which was playing when the remains were discovered.

The oldest known skeleton of an upright-walking hominid dates back 3.6 million years, 400,000 years before Lucy walked the earth. The five-foot-tall specimen, found in Ethiopia, was named Kadanuumuu, or "big man," by the locals.

IT'S THE PITS

Compact discs are made of polycarbonate. Data is encoded in the form of indentations, known as "pits." A scanning laser reads pits and a non-pits area as ones and zeroes.

HANDS-FREE

"Bluetooth" open wireless technology gets its name from King Harald Bluetooth of Denmark, who united the Danish tribes in the tenth century. Bluetooth, which was invented by Swedish telecommunications giant Ericsson, converts differing electronic communications methods into one universal standard, allowing cable-free connections between electronic devices.

BLOODY INTERESTING

In the early days of blood banks, donated blood was segregated according to race.

There are now blood banks for dogs and cats.

VERY PERSONAL HYGIENE

The bacterial population on each person is so individual that it may one day be used to establish identity.

IT'S A GAS

Five of the six naturally occurring "noble gases" are used in "neon" lights—argon glows lilac-purple, helium glows purple, krypton glows white, neon glows orange-red, and xenon glows blue. Only radon is not used.

Propane is stored as a liquid at concentrations near 250:1 of that of propane in the gaseous state.

GEMS IN A JIFFY

Man-made diamonds only take four days to produce.

Synthetic diamonds can be made that are identical to natural diamonds and can only be differentiated using spectroscopy tests.

Man-made diamonds retail for about one-third the price of natural diamonds.

"Memorial" diamonds are made using carbon taken from the hair or the ashes of the dead.

INFINITESIMAL INFORMATION

IBM engineer Don Eigler was the first person to move and control a single atom, in 1989, using an atomic force microscope. He later used the technique to spell out "IBM" with thirty-five xenon atoms.

Scientists can now even control the spin of a single electron.

An electron can be in more than one place at a time.

Positrons are antimatter versions of electrons.

The Fermi Gamma-Ray Space Telescope revealed recently that thunderstorms on Earth create antimatter.

EASY COME, EASY GO

The Permian impact event, also know as the Great Dying, is believed to have caused the greatest extinction in Earth's history, with 96 percent of marine life and 70 percent of land vertebrates wiped out. No one is sure what caused the Great Dying that occurred 251.4 million years ago, but it paved the way for dinosaurs to dominate the planet. They in turn were wiped out by the mass extinction of 65 million years ago.

Today's species make up only 1 percent of all the species that have ever existed.

GOOD DAY, SUNSHINE

Solar cells convert sunlight into electricity using the photovoltaic effect. Photons from sunlight knock electrons into higher states of energy, creating electric current. This can be used to charge batteries or run machinery.

The photovoltaic effect was discovered by French physicist A. E. Becquerel in 1839.

The first solar cell, made from the element selenium coated with gold, was constructed by American Charles Fritts in 1883.

One enriched uranium nuclear fuel pellet the size of the tip of a little finger contains the equivalent energy of 1,780 pounds of coal or 149 gallons of oil.

IT'S A BLAST

Each year, between fifty and seventy volcanoes erupt somewhere in the world.

In 2010, sixty volcanoes erupted somewhere around the world.

There are currently six active or potentially active (not erupting) volcanoes in California.

More than 80 percent of the earth's surface is of volcanic origin.

Lava is 1,300°F to 2,200°F when it first erupts from a volcano.

Lava is up to one hundred thousand times as viscous as water.

One-third of the lava to reach the surface of the earth since the Middle Ages is in Iceland.

When Mount St. Helens blew its top in 1980, the blast wiped out 230 square miles of forest.

The volcano that erupted in Iceland in 2010 and grounded flights in Europe is named Eyjafjallajökull, pronounced *AY-yah-fyah-lah-YOH-kuul*.

THE BIG "KRAK" UP

When the volcanic island of Krakatoa blew apart in 1883, the blast could be heard 3,500 miles away in Perth, Australia. It is believed to be the loudest noise produced in recorded history.

The explosion ruptured the eardrums of sailors in the nearby waters and the shock wave circled the globe seven times.

The force of the blast is estimated to be thirteen thousand times the strength of the atomic bomb dropped on Hiroshima.

The ash thrown into the atmosphere lowered the global temperature and affected weather patterns for five years.

TAKE NOTES

Doctors can now remove small organs, such as an appendix or gall bladder, through the mouth. The surgery, known as NOTES (natural orifice translumenal endoscopic surgery), involves passing tiny tools and a camera down the esophagus and through a tiny hole made in the stomach that does not require stitches. The same surgeries can be done through the vagina.

WEIGHTY MATTERS

Recent research shows that the more obese a postmenopausal woman becomes, the greater her decrease in memory, reasoning, and other mental skills.

Moderately obese drivers are 21 percent more likely to die in a car crash than normal weight individuals. Morbidly obese drivers are 56 percent more likely to die.

Close to 95 percent of people who lose weight dieting will later regain it.

RUNNING ON EMPTY

A gallon of jet fuel weighs 6.6 pounds.

A Boeing 767 burns 1,722 gallons of fuel an hour. That's about 190 pounds of fuel per minute.

To save money, airlines try to cut the weight of their planes by planning on having as little fuel left in the tanks when they land as possible.

SLOW BURN

Each year, one in one hundred thousand African Americans are diagnosed with melanoma, as are five in one hundred thousand Hispanics, nineteen in one hundred thousand white women, and twenty-nine in one hundred thousand white men.

NOTHING TO FEAR, BUT . . .

Alliumphobia is the fear of garlic.

Anthrophobia is the fear of flowers.

Dipsophobia is the fear of alcohol.

Medorthophobia is the fear of erections (in men, that is).

Oenophobia is the fear of wine.

Somniphobia is the fear of sleeping.

Sophophobia is the fear of learning.

FEAR NOT

A very rare condition, known as Urbach-Wiethe disease, renders sufferers unable to experience the feeling of fear.

Researchers tried to scare one woman with the disease using spiders, snakes, and haunted houses to no avail. The woman even reported no fear when she had previously been robbed at gunpoint.

GOING VIRAL

One virus can replicate itself one hundred times, meaning, in five generations, 10 trillion viruses can result from just one original.

ON ICE

Icebergs make a fizzing sound as they melt, which is known as "Bergie seltzer," which is caused by the popping of frozen air bubbles.

Icebergs are monitored by the U.S. National Ice Center.

The tallest icebergs can stick out of the water as high as a fifty-five-story building.

Some people camp on icebergs.

SHUTTLE SERVICE

The space shuttle has forty-four thrusters to help it maneuver in orbit. By firing these thrusters in different combinations, the shuttle can move in any direction.

The space shuttle orbiter uses a two-part fuel (fuel and oxidizer) that combusts when mixed. No ignition spark is needed.

SPEAK EASY

The position of the human voice box is lower in the throat than in other primates, giving humans a much larger resonating system with which to generate speech.

Because of this low voice box position, humans cannot swallow and breathe at the same time without choking, as other animals can.

The reason that infants can suckle and breathe at the same time is because their voice box does not drop until about the age of nine months.

TALKING DIRTY

According to a 2010 *Journal of Applied Microbiology* study, smartphones have more germs than a subway station toilet. Sharing the phones is just as likely to spread bacteria and viruses as sneezing in someone's face.

GET THE LEAD OUT

The symbol for lead is Pb, which is short for the Latin word for lead—*plumbum*. It is from this word that the English word "plumber" derives.

The Romans flavored food with lead, because it added sweet overtones. They also used the element to make water pipes. Both practices led to wide-spread lead poisoning in the Roman Empire.

Lead is the best substance to shield radiation. It is used in cathode ray tubes for this purpose.

Eighty percent of the world's lead goes into car batteries.

WATER, WATER, EVERYWHERE

There are 326 million trillion gallons of water on Earth.

The world's oceans contain 320 million cubic miles of water.

One trillion tons of water evaporate from the world's oceans each day.

The surface of the ocean is not uniform and flat. It has "hills" and "valleys," which reflect the contours of the seabed below. A seven-thousand-foot seamount will cause a three-foot "hill" on the surface.

There are 125,000 square miles of coral reefs on planet Earth.

There are no fish in the Great Salt Lake because it is too salty.

CHOOSE YOUR POISON

A 2010 British study ranked the most dangerous drugs, based on the harm they can do to the user, such as death, damage to mental functioning, and loss of relationships, and the harm they do to others. On a scale of 100, alcohol was ranked the most dangerous drug at 72 out of 100, followed by heroin (55), crack (54), crystal meth (33), cocaine (27), tobacco (26), speed (23), marijuana (20), benzodiazepines (like Valium) (15), ecstasy (9), anabolic steroids (9), LSD (7), and "magic" mushrooms (5).

> Alcohol kills approximately 2.5 million people a year when considered a main cause of heart and liver disease and car wrecks.

IT'S A MATTER OF TASTE

The little bumps visible on the tongue are not individual taste buds, but clusters of from fifty to one hundred taste buds, known as fungiform papillae.

> Any area of the tongue can detect any taste, although some areas are more sensitive to certain tastes.

The flavors of the foods a pregnant woman eats end up in the amniotic fluid and the fetus may acquire a preference for these tastes later in life.

> Thirty-five percent of people are sensitive to the smell of androstenone, an aromatic compound found in celery and pork, and claim it smells like stale urine.

Fifteen percent of people say androstenone smells woody or floral, and half of people can't smell it at all.

HOT SCIENCE

In 2010, the Large Hadron Collider particle accelerator at the CERN laboratory near Geneva, Switzerland, successfully smashed together lead ions traveling near the speed of light and produced a miniature "big bang" with subatomic fireballs with temperatures over ten trillion degrees Fahrenheit.

BREAKFAST BARFING

Morning sickness is known medically as hyperemesis gravidarum.

Morning sickness is theorized to be a way of protecting the fetus from food toxins.

Women who don't get morning sickness are more likely to have a miscarriage.

Women with sisters who had severe morning sickness are seventeen times more likely to develop it too.

🐿 EARLY ED

Infants probably start learning language in the womb. Three-day-old German babies cry with falling tones, like spoken German, while three-day-old French infants cry with a rising intonation, like spoken French.

THE COLD, HARD FACTS

Salt lowers the freezing point of water to 15°F (from 32°F).

Anything that dissolves in water will melt ice.

Airplane wings are deiced using warm glycol (antifreeze).

Shivering muscles create heat, which helps to warm the body.

Skin turns red when exposed to the cold (e.g., rosy cheeks) due to more blood being sent to the surface to warm the skin.

Soap bubbles will freeze if blown into air below –13°F.

THE DOCTOR IS IN

Greek doctor Georgios Papanikolaou invented the Pap smear.

An otolaryngologist is a doctor that specializes in the ears, nose, and throat.

The World Health Organization estimates that six hundred thousand people worldwide die each year as a result of exposure to secondhand smoke.

Five hundred thousand children a year worldwide go blind from vitamin D deficiency.

Kids who have both peanut and tree nut allergies are ten times more likely to be allergic to sesame seeds.

A recent survey found that people's stress levels drop markedly after the age of fifty.

New research suggests that drinking human breast milk may kill cancer cells.

Each year, 250 million people around the world contract malaria and almost 1 million die from the disease.

About 25 percent of the people on Earth host parasitic worms.

People with fibromyalgia can now enter a chamber chilled to −238°F for two to four minutes. This causes the body to release endorphins that relieve their pain.

BREATHLESS

Nearly 25 million Americans suffer from asthma.

Each year, 1.75 million people visit emergency rooms in the United States for asthma-related problems.

In 2008, asthma killed 3,395 Americans.

Females, children, blacks, Puerto Ricans, and poor people are the most likely to have asthma.

CURIOUS CONDITIONS

Reflex sympathetic dystrophy is a nerve disorder that causes severe burning pain in the limbs, as if the sufferer were on fire. It can be triggered by a stroke or sur-gery and has no cure, although treatment can lessen the symptoms.

Polyglandular Addison's disease is a hormonal disor-der wherein sufferers' bodies are not able to produce adrenaline in response to stress. Without adrenaline, the organs cannot respond to stress and go into shock and shut down. Sudden emotional stress can cause instantaneous death. Steroidal medications can make the condition manageable.

Men with post-orgasmic illness syndrome are allergic to their own semen. Sufferers may experience runny nose, burning eyes, feverishness, and extreme fatigue after ejaculation. These symptoms can last for up to a week. Hyposensitization therapy may help reduce its impact.

Babies are 33 percent more likely to die from sudden infant death syndrome (SIDS) on New Year's Day than any other day of the year. Researchers believe that parental use of alcohol may be a factor.

THAT SUCKS

In 2010, a woman in Australia suffered a stroke and was partially paralyzed from a hickey on her neck. Doctors determined that an amorous male had sucked so hard on the tissue above a major artery that it caused a blood clot to form that dislodged and traveled to her heart, causing a stroke.

THE EYE OF THE BEHOLDER

A recent scientific study found that 23 percent of over-weight women saw themselves as being smaller than they actually were. Eighty percent of overweight and obese African American women and 75 percent of Hispanic women had this misperception.

Thirteen percent of people weigh themselves every day, while 60 percent rarely or never do.

THEY ALL LOOK THE SAME

Most people experience the cross-race effect, wherein members of one race have more difficulty recognizing individuals from another race, compared to those within their race. It is a phenomenon documented in several studies.

CUBISM

There are 43,252,003,274,489,856,000 possible starting positions on a Rubik's Cube.

Even the most scrambled cube is never more than twenty moves away from being solved.

IT'S A WRAP

Originally, Saran Wrap had a greenish tint and an un-pleasant odor. The chemical it was made from was also sprayed on fighter planes and car upholstery to weather-proof them.

The shiny side of the aluminum foil gets that way from the highly polished steel rollers that make it.

SCIENCE SHORTS

The Bunsen burner was invented by Robert Bunsen's co-worker, instrument maker Peter Desaga, in 1855.

The Italians probably invented eyeglasses around 1250.

The fastest camera can take 1 million photographs per second.

The tires on a jetliner suffer the most strain while taxiing, not upon takeoffs and landings.

Bricks are made from clay and shale with trace amounts of barium added so they don't discolor when fired.

Cloth diapers are no more Earth-friendly than disposables. Washing one baby's cloth diapers will use 22,500 gallons of water, before they are potty trained.

The first MP3 player only held ten songs.

There is three feet of DNA in every cell.

Bar codes were first used by the railroads to keep track of cars.

SPACED OUT

SUNNY SIDE UP

The sun is 27 million degrees Fahrenheit at its core.

Sunspots are caused by "kinks" in the sun's magnetic field, which keep the hot gases from rising to the surface. Sunspots are 2,000°F cooler than the surrounding surface.

The most massive known star is R136a1. Weighing 265 times as much as the sun, it shines 10 million times brighter.

MANY MOONS

There are 168 moons orbiting six of the eight planets.

Saturn has forty-seven moons.

Mercury and Venus have no moons.

None of the solar system's known moons have moons of their own.

Jupiter's ice-covered moon Europa is the smoothest body in the solar system.

> There is thought to be a liquid water ocean beneath Europa's ice that may be up to sixty-two miles deep.

Saturn's moon Titan is bigger than the planet Mercury and 50 percent bigger than Earth's moon.

> Titan's surface has the consistency of pudding.

Enceladus is another moon of Saturn. It has liquid geysers that shoot plumes of what may be water hundreds of miles into space.

> The smelliest place in the solar system is probably Jupiter's moon Io, which has an atmosphere believed to stink like rotten eggs from the hydrogen sulfide its volcanoes constantly spew.

Without the moon's stabilizing influence, the earth would rotate at twice its present speed, creating enormous tides and winds in excess of two hundred miles per hour.

> The moon is collapsing. In the past 1 billion years, the moon has shrunk several hundred feet in radius.

Earthshine is the faint illumination of the dark part of the moon's disk produced by the reflection of sunlight from the earth.

Neil Armstrong placed mirrors on the moon's surface that researchers use to bounce back laser beams shot from Earth that allow them to accurately measure the distance to the moon.

DAILY PLANET

All the planets in the solar system orbit in the same direction as the sun's rotation.

Jupiter's "Great Red Spot" is a massive storm three times bigger in circumference than Earth. It has been raging since the Middle Ages. Neptune also has a massive storm known as the "Great Dark Spot."

There is a huge, near perfectly hexagonal-shaped storm raging at Saturn's north pole that measures 13,800 kilometers on each of its six sides. At 25,000 kilometers wide, it is large enough to fit four Earths inside.

The atmospheric pressure on the surface of Venus is ninety-three times that of Earth. This pressure is comparable to being one kilometer deep in the ocean.

Due to its high concentrations of atmospheric carbon dioxide and sulfur dioxide, Venus has the greatest greenhouse effect of any planet.

Valles Marineris is an enormous valley on Mars that is up to 6.2 miles deep and 2,485 miles long. (By comparison, the Grand Canyon is only a little over a mile deep.)

The atmospheric pressure at the surface of Mars is just 0.0006 times that found on Earth.

Jupiter's gravity attracts much of the debris in the outer solar system that otherwise might crash into Earth.

Like Earth, Jupiter and Saturn both have auroras at their poles.

LORD OF THE RINGS

Saturn's rings are as little as ten feet thick in some places, but average about sixty feet.

The rings are made up of ice particles ranging in size from microscopic to the size of a house.

Saturn's rings reflect sunlight onto the dark side of the planet. This is known as "ringshine."

The gaps between Saturn's rings are caused by the gravitational forces exerted by its sixty-two moons.

Like Saturn, Jupiter has rings. They are comprised of dark dust and don't reflect sunlight as well as Saturn's icy rings.

The planet Uranus also has rings.

A GALAXY FAR, FAR AWAY

The Milky Way got its name from Greek mythology—the goddess Hera spilled her breast milk when refusing to nurse Heracles (Hercules), forming the Milky Way.

> The density of stars at the center of the Milky Way galaxy is 1 billion times greater than in the region of our solar system.

The Milky Way and nearby Andromeda galaxies are moving toward each other at a speed of 1 million miles per hour. They will collide in 4 billion years. Since there are trillions of miles between the stars in each galaxy, there probably won't be any actual collisions, but simply a merging of the two galaxies.

> For a brief time, a supernova outshines the galaxy it is in.

BRILLIANT BURSTS

Short gamma ray bursts are thought to result from the collision of two neutron stars. They produce more energy in a fraction of a second than the sun will produce in 100 million years.

> Long gamma ray bursts are the result of collapsing giant stars. They release more energy in several seconds than the sun will produce in its 10-billion-year life span.

THE FOUR DWARFS

Brown dwarfs, which are celestial objects between the size of a giant planet and a small star that emit infrared radiation, used to be called black dwarfs until 1975.

Brown dwarfs are not really brown, but dull red.

Today, black dwarfs are defined as white dwarfs that have cooled down to the point where they emit no significant heat or light. At present, they remain theoretical, as the time required for them to cool down to that state is longer than the current age of the universe.

A white dwarf is a very small, dense star about the size of the earth. They are believed to be the remnant cores of very old stars that have shed their outer layers.

Red dwarfs have a mass less than half that of the sun. They are cool and emit little light. Even though none are visible to the naked eye, they are the most common type of star in the Milky Way galaxy.

STARRY, STARRY NIGHT

New scientific estimates put the number of stars in the universe at 300 sextillion. That's 300,000,000,000,000,000, 000,000.

WHAT'S THE MATTER?

Approximately 5 percent of the universe is made up of ordinary matter, that is, matter that scientists can physically observe and is made up of atoms. Dark energy makes up about 73 percent and the rest consists of dark matter.

Dark matter is named such, not because it is dark, but because scientists don't know what it is.

Scientists think that there is something huge beyond the edge of the visible universe that is pulling on its galaxies. They call this motion "dark flow."

PEE WEE PLANETS

The asteroid belt that lies between the orbits of Mars and Jupiter is comprised of mainly rocky objects, while the Kuiper Belt at the edge of the solar system is mainly composed of icy bodies.

Pluto used to be classified as a planet, but now is considered a Kuiper Belt object.

There are seventy thousand Kuiper Belt objects.

The dwarf planet Haumea, in the Kuiper Belt, is the fastest spinning object in the solar system, rotating once every four hours.

The sun's gravitation is believed to extend to the hypothesized Oort cloud of comets that lies roughly a light year away.

Scientists estimate that the Oort cloud contains some 400 billion objects.

Ninety percent of the material in the Oort cloud is believed to have come from stars other than the sun.

OUT-OF-THIS-WORLD ACCOMMODATIONS

It takes the International Space Station (ISS) ninety minutes to make one trip around the globe and it travels at a speed of 17,239 miles per hour.

The ISS is large enough to be seen from Earth with the naked eye.

The ISS has been continually manned since 2000.

With an estimated final cost of more than $150 billion, it is the most expensive object ever created.

FASTER THAN A SPEEDING BULLET

The fastest any humans have ever traveled was the astronauts on *Apollo 10*, who zipped along at twenty-five thousand miles per hour.

Apollo missions 18, 19, and 20 were all canceled due to budgetary reasons.

LONG DISTANCE VOYAGER

The *Voyager 1* spacecraft, launched in 1977, is still traveling at thirty-nine thousand miles per hour outside of the solar system, more than 10 billion miles from Earth.

It records its scientific data on 8-track tape machines.

It takes twelve hours for *Voyager*'s signals to make it back to Earth.

STATE OF THE UNION

GARDEN STATE

The world's first boardwalk was built in Atlantic City, New Jersey, to help keep the beach sand out of hotel lobbies in 1870. Alexander Boardman came up with the idea and it was called "Boardman's Walk," later shortened to Boardwalk. The Atlantic City Boardwalk stretches eight miles.

There are more horses per square mile in New Jersey than in any other state.

The northwestern counties of New Jersey have the highest concentration of black bears per square mile than any other area in North America.

Fort Lee, New Jersey, was the center of America's first movie industry.

TAR HEEL STATE

America's first gold rush took place near Charlotte, North Carolina, in the early 1800s.

North Carolina had the most casualties of any Confederate state during the Civil War.

LITTLE RHODY

Rhode Island is the only state where V-J (Victory over Japan) Day is an official state holiday. It is observed on the second Monday of August.

Rhode Island has the highest concentration of Catholics in the country.

Having a low license plate number is a status symbol in Rhode Island. Some residents will pay thousands of dollars to buy someone's low digit plate and people leave them to relatives in their wills.

SILVER STATE

Nevada is the most arid state.

Nevada is the third-largest gold mining district in the world (behind South Africa and Australia).

The Comstock Lode, in Nevada, is the largest silver deposit ever found anywhere.

Nevada has the most mountain ranges of any state.

Nevada outlawed gambling in 1910, but made it legal again in 1931.

Las Vegas has the most hotel rooms of any city in the world.

Las Vegas is the most populous American city founded in the twentieth century. (Chicago was the most populous city founded in the nineteenth century.)

Las Vegas was named by the Spanish for the green "meadows" (*vegas* in Spanish) that once grew in the area.

OLD DOMINION

Virginia's nickname—Old Dominion—comes from King Charles II of England, who named the colony "Dominion" upon his restoration to the throne in 1660, for its loyalty to him during the English Civil War.

The slogan "Virginia Is for Lovers" was originally "Virginia Is for History Lovers."

THE LAST FRONTIER

Alaska used to be known as Russian America, the Department of Alaska, and the District of Alaska.

When Alaska was made a state in 1959, it increased the size of America by one-fifth.

The largest national park in America is Wrangell–St. Elias in Alaska at 13.2 million acres. The next four largest national parks are also in Alaska.

PINE TREE STATE

Someone from Maine is known as a Mainer.

The Damariscotta Pumpkinfest & Regatta in Damariscotta, Maine, has an annual race where participants paddle in enormous hollowed out pumpkins.

Maine has the highest median age of any state—42.2 years—meaning half the residents are younger than this and half are older.

The biggest "snowman" ever was built by the residents of Bethel, Maine, in 2008. The 122-foot, 1-inch snowwoman didn't entirely melt until July.

BELLWETHER STATE

The state of Missouri has voted for the winner in every presidential election since 1904, except in 1956 (Missouri voted Adlai Stevenson) and 2008 (John McCain).

NORSE DAKOTA

The early white settlers of the Dakotas were primarily from Norway and Iceland.

North Dakota and South Dakota joined the Union on the same day—November 2, 1889. However, no one recorded which state was admitted first, so North Dakota, which comes alphabetically before South Dakota, is generally listed as the thirty-ninth state and South Dakota as the fortieth state.

BUCKEYE STATE

The name "Ohio" derives from the Iroquois Indian word for "big river."

The Ohio buckeye tree is also known as the fetid buckeye because of the offensive smell that is generated from the flowers, wood, and crushed leaves of the species.

Ohio used to be known as the Northwest Territory.

EMPIRE STATE OF THE SOUTH

Georgia fancies itself as the "Empire State of the South."

Sweet Vidalia onions are only grown in south Georgia.

ROCKY MOUNTAIN STATE

Colorado is the highest state, with fifty-four mountains above fourteen thousand feet.

Denver International Airport, at fifty-three square miles, is twice the size of Manhattan.

BADGER STATE

Wisconsin is the Badger State. The nickname comes from early miners who settled the area and burrowed homes into the side of hills, much like a badger.

The National Mustard Museum, in Middleton, Wisconsin, has more than four thousand kinds of mustard.

FLAT WATER STATE

The name "Nebraska" comes from the Omaha Indian word meaning "flat water," for the Platte River which runs through the state.

Nebraska is the only state that has a nonpartisan, unicameral legislature, meaning the legislators are elected with no party affiliation next to their names and a one-house legislature.

Ninety-three percent of the land in Nebraska is devoted to agriculture.

TAX HAVEN

A person from Delaware is known as a Delawarean.

Delaware used to be known as the Lower Counties on the Delaware and was owned by William Penn, who also owned adjacent Pennsylvania.

Sixty percent of Fortune 500 companies are incorporated in Delaware because of its business-friendly corporate tax and liability laws.

> The first Thursday after an election is "Return Day" in Delaware. Historically, the candidates and citizens would gather in Georgetown, Delaware, on this day to hear the vote tallies. The tradition of the winners and losers riding together in a horse-drawn carriage to the town square for a ceremonial burying a hatchet in a box of sand continues today.

LAND OF ENCHANTMENT

Santa Fe, New Mexico, is the oldest capital city in North America, established 1610.

> Santa Fe means "holy faith" in Spanish.

Roughly 29 percent of New Mexicans speak Spanish at home, according to the 2000 Census.

> New Mexico has the most PhDs per capita of any state.

Acoma Pueblo, also known as Sky City, is a pueblo built atop a 367-foot mesa in New Mexico that is one of the oldest continuously inhabited settlements in the United States, dating back nearly one thousand years. The Hopi village of Old Oraibi in Arizona is likewise about one thousand years old.

LEFT COAST

California gets its name from the mythical island of California, mentioned in the 1510 fictional book *Las Sergas de Esplandián*, which was inhabited only by black Amazon women and ruled by Queen Califia.

The world's first motel—the Milestone Motel—was located in San Luis Obispo, California. Today, it is known as the Motel Inn of San Luis Obispo. This location was picked because it was halfway between Los Angeles and San Francisco, which were separated by a two-day car trip.

There are sixteen thousand windmills in California.

California's Mount Whitney is the highest point in the continental United States

One-half of the fruits and vegetables grown in the United States come from California.

Grapes are California's number one cash crop.

LONE STAR STATE

Before gaining independence from Mexico, Texas was part of the Mexican state Coahuila y Texas, which had a two star flag. When Texas broke away, it kept one of the stars on its new flag and later became the Lone Star State.

Two of the reasons the Republic of Texas joined the United States was because of its massive debt and an inability to defend itself from repeated Mexican incursions.

EMPIRE STATE

Niagara Falls State Park, in New York, is the nation's first and oldest state park.

BEAVER STATE

Oregon leads the nation in timber production.

BIJOU STATE

Forty percent of the country's coastal wetlands are in Louisiana.

The busiest port in America is the Port of South Louisiana. The port runs fifty-four miles along the Mississippi River from New Orleans to Baton Rouge, Louisiana.

GREEN MOUNTAIN STATE

Vermont was the first state to outlaw slavery.

BLUEGRASS STATE

Kentucky was a slave state that did not secede from the Union during the Civil War.

Kentucky Colonel is an honorary title bestowed by the governor and secretary of state of Kentucky.

The most recognized Kentucky Colonel is Harland Davis Sanders, also known as Colonel Sanders, the founder of KFC (formally known as Kentucky Fried Chicken). Other honorees include Elvis Presley, Muhammad Ali, Winston Churchill, Alton Brown, Johnny Depp, Babe Ruth, Richard Petty, and Tiger Woods.

VOLUNTEER STATE

Ninety-one of Tennessee's ninety-five counties are dry.

Jack Daniel's whiskey is distilled in Moore County, Tennessee, which is a dry county.

BEEHIVE STATE

While Tennessee may bill itself as the Volunteer State, the state where the greatest percentage of the population does volunteer work of some kind is Utah, with 45 percent.

AMERICANA

STATE OF DECLINE

From 2008 to 2009, the population decreased in just three states—Maine, Michigan, and Rhode Island.

The states with the oldest median age are Maine, Vermont, West Virginia, New Hampshire, and Florida, in that order.

The states with the youngest median age are Utah, Alaska, and Texas.

SPRING AHEAD

Twenty-seven percent of Americans say that they have been either early or late for work because they forgot to change their clocks for daylight saving time.

In 1965, Minneapolis and St. Paul did not begin daylight saving time on the same day, resulting in the two adjacent cities being on different time.

Most of the world's countries do not use daylight saving time. It is primarily used in the Northern Hemisphere.

In the United States, Arizona and Hawaii do not observe daylight saving time, although the Navajo Nation within Arizona does so.

UNIFORM CODE

In the United States, 17.5 percent of public schools require uniforms.

Twelve percent of American high schools produce fifty percent of the nation's dropouts.

AFFORDABLE HEALTH CARE?

In 2010, the United States spent $7,285 per capita on health care.

The United States spent more on health care in 2009 than the total economic output of the United Kingdom, Russia, France, and Brazil.

Sixty-nine percent of the drugs dispensed in the United States are generic, but generics only account for 19 percent of prescription spending.

CAT PEOPLE

Seventy-four percent of Americans say they like dogs a lot, as opposed to 41 percent who like cats a lot, according to an Associated Press survey. The poll also found that 15 percent dislike cats a lot, while only 2 percent dislike dogs a lot. Twenty-eight percent of Americans would give CPR to a dog or cat.

OUT-OF-TOWNERS

Miami has the highest percentage of foreign-born residents of any American city—36.5.

Chicago has the largest ethnically Polish population, after Warsaw.

BIG APPLES

New York City was first called the "Big Apple" by John J. Fitz Gerald, a sportswriter for the *New York Morning Telegraph*, in 1921.

Other cities have taken similar nicknames. Manhattan, Kansas, is the "Little Apple." Atlanta is the "Big Peach." Tampa is the "Big Guava." Sacramento is the "Big Tomato." And Omaha is known as the "Big O."

THE CITY THAT NEVER SLEEPS

In 1902, there were sixty-five skyscrapers under construction in New York City.

There are fourteen thousand miles of steel cable holding up the Brooklyn Bridge.

New York has more Jews than Tel Aviv.

New York City has the most expensive retail space in the world. It costs an average of $1,725 per square foot per year. Sydney, Hong Kong, London, and Paris round out the top five.

The Bronx is the only New York City borough that is not on an island.

POWER UP

Vermont gets 72 percent of its electricity from nuclear power, New Jersey gets 55 percent, Connecticut gets 53 percent, South Carolina gets 52 percent, and Illinois 49 percent.

Wyoming is the state that uses the most energy per capita. New York uses the least.

MELTING MOUNTAINS

At one time the Appalachians were as tall as the Himalayas.

HOT SPOTS

Yellowstone National Park contains 70 percent of the world's geysers.

DEEP THOUGHTS

Four hundred people a year are rescued by helicopter from the Grand Canyon.

It took the Colorado River 5 million years to carve the Grand Canyon.

In 1907, 3,242 American mine workers died on the job. That number had fallen to thirty-four in 2009. Most died from aboveground truck and machinery accidents.

TALL TALES

Sears Tower in Chicago is now known as Willis Tower. Willis Group Holdings leased part of the building in 2009 and got to rename the skyscraper. Sears sold the building in 1994.

> Willis Tower is the tallest building in the Americas and the fifth tallest in the world.

In 2009, the building opened three glass observation boxes that protrude out from the structure about four feet. Tourists can step out onto the glass floor and look straight down 103 floors (1,353 feet).

> Willis Tower has sixteen thousand windows.

The Art Deco Chrysler Building in New York City was the tallest skyscraper in the world for eleven months, until it was surpassed by the Empire State Building in 1931.

> The Chrysler Building was built with the personal money of Walter P. Chrysler in 1929–30 and served as the automaker's headquarters until the 1950s.

The Chrysler Building's design includes many elements from Chrysler cars of the time. There are four chrome eagles at the corners of the 61st floor that are replicas of 1929 Chrysler hood ornaments and four chrome replicas of 1929 Chrysler radiator caps adorn the 31st floor.

CAUSE OF THE MONTH CLUB

In America, each month has been tapped to honor some very curious things. Here are a few examples:

January is National Jump Out of Bed Month.

February is Return Shopping Carts to the Supermarket Month.

March is National Frozen Foods Month.

April is National Humor Month.

May is National Barbecue Month.

June is Potty Training Awareness Month.

July is Cell Phone Courtesy Month.

August is Happiness Happens Month.

September is National Biscuit Month.

October is National Popcorn Poppin' Month.

November is Peanut Butter Lover's Month.

December is National Tie Month.

IS THAT LEGAL?

In Arkansas, Maryland, Pennsylvania, South Carolina, Tennessee, and Texas it is illegal to hold public office if you don't believe in God. (Obviously, these laws are rarely enforced.)

The City of Detroit has legally forbidden all employees from wearing perfume, cologne, or any scented deodorant, lotion, gel, hairspray, etc., on the job.

In 2005, law enforcement officials in the United States conducted 1,800 wiretaps. These do not include wiretaps used by intelligence gathering agencies.

A 2008 study found that the most dangerous cities in America were Camden, New Jersey; Detroit, Michigan; St. Louis, Missouri; Oakland, California; and Flint, Michigan, in that order.

Phoenix, Arizona, has the dubious distinction of being the "kidnapping capital" of America. In 2008, there were more than 370 reported kidnappings, second in the world only to Mexico City.

July and August are the months with the highest rate of violent crime.

In 2009, New York City had 471 murders.

Between 1990 and 2009, homicides in New York City fell 79 percent. The number of homicides in Los Angeles fell by 68 percent and Chicago's fell 46 percent during the same time period.

As of the end of 2010, thirteen states had decriminalized marijuana possession, meaning that those caught with small amounts are merely fined and the incident does not become part of the offenders' criminal record.

The more liquor stores there are in an area, the greater the rate of domestic violence in that area.

In 2010, there were approximately 1 million gang members in the United States. Chicago had 105,000 of them.

Each year 12 million Americans have their mug shots taken.

In 2010, there were forty-six executions in the United States, seventeen of them in Texas.

At the end of 2010, there were 697 inmates on death row in America.

A 2010 study by the U.S. Forest Service found that neighborhoods with big trees had less crime than those with small trees.

HALLOWED HALLS

Harvard University is officially known as the President and Fellows of Harvard College.

Harvard was the first corporation chartered in what would become the United States.

When the school first opened, in the late 1630s, students could pay their tuition with things like cloth, boots, hardware, livestock, and agricultural products.

FIRST FACTS

Grover Cleveland was the first president captured on film, signing a bill in 1895.

Woodrow Wilson was the first American president to hold a press conference, in 1913.

Ronald Reagan was the first U.S. president to use email.

George W. Bush is the only president to have graduated from both Yale and Harvard.

At age eighteen, George H. W. Bush enlisted in the military when Pearl Harbor was attacked and became the youngest Navy pilot of the time.

Barack Obama and Sarah Palin are tenth cousins through common ancestor John Smith. Obama and Rush Limbaugh are tenth cousins once removed.

JFK took painkillers for his back pain every day while president.

Ronald Reagan acquired his nickname "Dutch" after he was born and his father commented that he looked like a "fat little Dutchman."

Richard Nixon picked up the nickname "Tricky Dick" from a 1950 Democratic ad leading up to his run for the California U.S. Senate seat.

◖ INITIALLY YOURS

The "S" in Ulysses S. Grant means nothing. He was born Hiram Ulysses Grant, but when his congressman nominated him to West Point as a teenager, he accidentally did so as Ulysses "S." Grant. Grant decided to keep the "S" because he liked the sound it.

The "K" in James K. Polk stands for Knox.

The "B" in Rutherford B. Hayes stands for Birchard.

The "A" in James A. Garfield stands for Abram.

The "A" in Chester A. Arthur stands for Alan.

The "G" in Warren G. Harding stands for Gamaliel.

LET'S TALK TURKEY

President John F. Kennedy allowed a Thanksgiving turkey he was presented with in 1963 to live, but it was President George H. W. Bush who formalized the "pardoning" of a Thanksgiving turkey in 1989.

Actually, one turkey and one alternate are "pardoned." In 2008, the alternate was used in the ceremony, as the first choice turkey fell ill.

Pardoned turkeys are often sent to Disney World or Disneyland to live out their days. In 2010, President Obama sent a bird named Apple and his alternate Cider to live at the Mt. Vernon estate of George Washington.

WORLD'S MOST DANGEROUS JOB

One in four U.S. presidents has been attacked by assassins.

One in ten U.S. presidents has been assassinated.

IN THE LINE OF FIRE

The armored presidential limo is known as the "Beast." It has special foam in the gas tanks to extinguish a fire instantly.

The president always travels with his own emergency blood supply.

The Secret Service uses specially trained Belgian Malinois dogs to sniff out bombs.

The Secret Service forensics lab has eight thousand different ink samples to help trace threatening letters written to the president.

The Secret Service agents who guard the president must qualify on several weapons every month.

George W. Bush was code-named "Tumbleweed" and "Trailblazer" by the Secret Service.

The annual Secret Service budget is roughly $1.5 billion.

In 1950, a gun battle erupted outside of Blair House in Washington, DC, where President Truman was living during the renovation of the White House. Two Puerto Rican nationalists tried to shoot their way in. One was killed by White House police officer Leslie Coffelt, who was killed by the attackers, and the other wounded.

The Bethesda Naval Hospital in Maryland has a special armored room available to the president for treatment in the event of a biological, chemical, or nuclear attack.

MILLIONAIRES CLUB

The five richest members of Congress in 2009 were:

Senator John Kerry (D-MA): $188.6 million

Rep. Darrel Issa (R-CA): $160.1 million

Rep. Jane Harmon (D-CA): $152.3 million

Sen. Jay Rockefeller (D-WV): $83.7 million

Rep. Michael McCaul (R-TX): $73.8 million

They all increased their wealth that year, despite the severe recession hitting the country.

Nearly one-half of members of Congress have assets of over $1 million.

The median worth of a U.S. senator was $2.38 million in 2010.

DISTRICT DATA

In 1789, there were sixty-five U.S. congressional districts. The number has since increased to 435.

New York's 15th District, which includes Upper Manhattan and Rikers Island, is the smallest district by area—10.29 square miles.

Montana has one congressional district. It is the largest congressional district by population, representing more than nine hundred thousand constituents. Alaska also has one congressional district, with the largest land area and lowest population density of any U.S. district. Delaware, Vermont, North Dakota, South Dakota, and Wyoming also have just one congressional district.

California has fifty-three congressional districts.

The oldest congressional district is the Delaware at-large district, which has had the same boundaries since 1787.

Four states—Delaware, Iowa, Mississippi, and Vermont—have never elected a woman to the House of Representatives.

KEEPING IT IN THE FAMILY

There has always been a member of the Dingell family representing Michigan's 15th Congressional District. The district was created in 1933 and John Dingell Sr. was elected in that year as the district's first member of the House of Representatives. Upon his death in 1955, his son John Dingell Jr. took over his seat and still held it as of 2011.

WHAT A WASTE

The United States Department of Energy Waste Isolation Pilot Plant in New Mexico stores nuclear waste in tunnels excavated in salt deposits 2,150 feet below the surface. This nuclear material that is created by the Department of Defense will remain hazardous and cannot be disturbed for ten thousand years.

THE STRAIGHT DOPE

Western states have straight borders because they had very few rivers to use as boundary lines and the railroads running through these states generally followed straight lines. Straight lines were also easier for the politicos back in Washington to draw on a map.

URGENT NEWS

Each day, 833 Americans go to the emergency room because of dog bites and 23 are admitted for their injuries.

The average cost of a dog bite hospitalization is $18,200.

Two hundred thousand Americans go to the ER each year for food allergy symptoms. That includes ninety thousand visits for anaphylaxis.

Every year, 1.5 million Americans visit emergency rooms due to cold symptoms.

Almost twelve thousand American children go to the ER each year because of poisoning related to household cleaning products.

SLEEP TIGHT

The U.S. cities with the largest bedbug infestations in 2010 were Cincinnati and Columbus, Ohio; Chicago, Illinois; Denver, Colorado; and Detroit, Michigan.

Howard Stern and Renée Zellweger have both had bedbug infestations in their homes.

AFFLUENT ADDRESSES

In 2010, the most expensive zip code in the United States was 91008 (Duarte, California), with a median household income of $4,276,462. The next top four were 94027 (Atherton, California), 90274 (Rolling Hills, California), 07620 (Alpine, New Jersey), and 10014 (New York, New York).

PLOWED UNDER

Between 1992 and 2007, 1,538 American farm workers died as a result of tractor overturns.

WHEEZIN'

Los Angeles is the American city with the greatest amount of ozone pollution. L.A. also ranks third in year-round particle pollution.

LOCAL WEATHER UPDATE

Hail causes approximately $1 billion in damage to crops and property each year in the United States.

The hottest temperature ever recorded in North Dakota was 121°F. The hottest temperature ever recorded in Florida was 109°F.

Minnesota was the state with the most tornadoes in 2010 with 104.

San Francisco had two and a half inches of snow on Christmas Day 1856.

The coldest temperature ever recorded in the United States was −80°F at Prospect Creek Camp along the Alaskan pipeline in 1971.

Summer temperatures in Fairbanks, Alaska, occasionally reach into the 90s.

Mount Baker, Washington, holds the U.S. record for most snow in one year—ninety-five feet—during the winter of 1998–99.

ABOVE AND BEYOND

The Medal of Honor is the highest U.S. military decoration. Only ten have been awarded since the end of the Vietnam War.

More than half of the Medals of Honor awarded since 1941 have been done posthumously.

Each branch of U.S. armed forces has its own Medal of Honor, except the coast guard and marines, who use the navy medal.

Only one member of the coast guard has ever received the Medal of Honor. Only one woman has won the medal.

Nineteen service members have won more than one Medal of Honor, fourteen of which were awarded for separate actions.

In total, there have been 3,470 medals awarded—including 1,522 in the Civil War, 124 in World War II, 133 in the Korean War, 246 in the Vietnam War, 4 in the Iraq War, and 4 in the Afghanistan War.

The most decorated American soldier of World War II was Second Lieutenant Audie Murphy, who single-handedly repelled five German tanks and infantry while also directing U.S. artillery fire, even after being wounded in the leg. In total, Murphy received thirty-three American and foreign medals and citations for his heroism throughout his twenty-seven-month tour of duty in Europe. After the war's end, Murphy became a movie star, acting in forty-four films.

THE LIGHT BRIGADE

The maximum allowable weight in the U.S. Army for a five-foot-eight-inch woman is 150–164 pounds. The maximum allowable weight for a five-foot-eight-inch man is 170–180 pounds.

One in four American high school seniors cannot pass the United States Army entrance exam.

BURN NOTICE

Eighteen percent of white American women and 6.5 percent of white American men use tanning beds. Most don't know that using tanning beds can increase their risk of skin cancer.

Each year, 2 million cases of skin cancer are diagnosed in the United States. That's more than all the cases of breast, colon, lung, ovarian, pancreatic, prostate, and uterine cancers combined.

NEWS FLASH

Seventy percent of American women have hot flashes at some point in their lives. Eighty-two percent of black women get them, as opposed to 68 percent of whites and Hispanics. Only about 59 percent of Japanese and Chinese women experience them.

MINOR PROBLEM

Of the 840,279 missing person cases in 2001, almost 800,000 were for minors. About 100 were children abducted by strangers. The vast majority of missing juveniles were runaways or were taken by family members involved in custody disputes.

DAY AT THE PARK

Each year 5 million people visit the San Diego Zoo, 3.5 million visit Yosemite National Park, 2.5 million visit the Alamo, and 400,000 visit the Rock and Roll Hall of Fame and Museum in Cleveland, Ohio.

The Blue Ridge Parkway in North Carolina and Virginia is the most popular unit of the National Park System, with more than 15 million annual visitors. The next four most popular National Park System properties are Golden Gate National Recreation Area, Great Smoky Mountain National Park, Gateway National Recreation Area in New Jersey/New York, and Lake Mead National Recreation Area in Arizona/Nevada.

THANK YOU FOR SMOKING

During the first 150 years of the American Colonies, tobacco was the number one crop export.

Americans spent $87.9 billion on tobacco products in 2009. Nearly 24 percent of Americans smoke cigarettes.

Eighty percent of African American smokers prefer menthol cigarettes, compared with 30 percent of Hispanic Americans and 22 percent of white Americans.

Menthol cigarettes make up 30 percent of the cigarette market.

WATER OVER THE DAM

Lake Mead, which was created by Hoover Dam, is the largest reservoir in the United States. It's named for Elwood Mead, who oversaw its construction.

The Ogallala aquifer under the Great Plains is the largest in the nation.

CAN DIAL-UP INTERNET BE FAR BEHIND?

Tiny Iowa Hill, California (population 200), which is fifty-eight miles northeast of Sacramento, finally got landline telephone service for the first time in 2010.

CLOAK AND DAGGER

There are sixteen different intelligence agencies in the United States, with 1,271 government organizations and 1,931 private companies also involved in intelligence gathering.

There are 854,000 Americans with a top secret clearance rating.

The United States intelligence budget for 2010 was $50 billion.

The National Reconnaissance Office (NRO) was set up in 1960 to build, launch, and operate spy satellites. The NRO was only revealed to the public, by the *New York Times*, in 1985.

THE RIGHT TO BEAR ARMS, OR NOT

The states with the most restrictive gun laws are California, Connecticut, Maryland, Massachusetts, New Jersey, and New York.

The states with the least restrictive gun laws are Alaska, Arizona, Idaho, Kentucky, Oklahoma, and Utah.

DRY NIAGARA

In 1969, the water flowing over the American side of Niagara Falls was shut off to allow cleaning of the riverbed and removal of loose rocks from the face of the falls. Two bodies were found in the rocks at the falls' base.

Only 10 percent of the water at Niagara Falls flows over the American side.

THE STROKE BELT

The region of the country including Tennessee, North and South Carolina, Georgia, Alabama, Mississippi,

Louisiana, and Arkansas has been called the "Stroke Belt." These states have a much higher rate of stroke deaths than the rest of the nation.

CALLING OUT

American workers took an average of fourteen sick days in 2007.

> Workers aged fifty-five to sixty-four took an average of eighteen sick days, compared with ten taken by workers aged sixteen to twenty-four.

The higher a family's income, the less likely that their children will miss days of school.

HOT WHEELS

The most popular vehicle for car thieves in the United States is the Cadillac Escalade, followed by the Ford F-250 crew cab pickup truck, the Infiniti G37, the Dodge Charger with the Hemi engine, and the Chevy Corvette Z06.

> The cars least likely to be stolen are the Saturn VUE and the Nissan Murano.

More cars are stolen in the United States on New Year's Day than any day of the year. Halloween is the second most popular day to steal cars. Car thieves apparently take off Christmas Day. It is the day the least number of vehicles are taken.

COLOR MY WORLD

The most popular color for luxury cars in the United States is black. The most popular color for compact/sports cars is silver. The most popular color for light trucks is white/white pearl.

CLICK IT OR TICKET?

Thirty-nine percent of cops killed in car crashes were not wearing a seat belt and 24 percent of them were ejected from the vehicle as a result.

Forty-two percent of police deaths behind the wheel involve single car crashes with stationary off-road objects.

Almost two out of three teens killed as occupants of motor vehicles are unbelted.

SPEED DEMONS

In 2005, 38 percent of male drivers aged fifteen to twenty who were involved in fatal crashes were speeding.

In 2005, 83 percent of traffic deaths resulting from speeding were not on interstate highways.

According to a 2010 National Highway Transportation Safety Report, one-third of all drivers killed in accidents tested positive for some kind of legal or illegal drug.

DRIVEN TO DISTRACTION

A 2010 study of states that outlawed texting while driving found that banning the practice actually resulted in a slight increase in accidents. This may be because drivers in those states hold their phones lower to avoid being seen texting and have even less eye contact with the road.

"Only" 18 percent of fatal crashes involving distracted drivers were the result of cell phone usage. The rest were caused by other distractions.

CRASH TEST DUMMIES

The Hyundai Sonata and BMW 5 series were the only cars to receive a five-star rating in the National Highway Traffic Safety Administration's test crashes of 2011 cars. All other models got four stars, except the Toyota Camry, which earned three stars, and the Nissan Versa with just two.

In 2009, 33,808 people died on America's roadways.

According to Allstate Insurance Company, the cities with the safest drivers are Fort Collins, Colorado; Chattanooga, Tennessee; Boise, Idaho; Colorado Springs, Colorado; and Knoxville, Tennessee, in that order.

The cities with the most dangerous drivers are Washington, DC; Baltimore, Maryland; Glendale, California; Newark, New Jersey; and Hartford, Connecticut, in that order.

JAMMIN'

The cities where drivers spend the most time stuck in traffic are Los Angeles, Washington, DC, and Atlanta, in that order.

IN GOOD HANDS

The states with the cheapest car insurance are Massachusetts, New Hampshire, Vermont, Minnesota, and Utah.

The states with the highest car insurance rates are Louisiana, Oklahoma, Mississippi, New York, and Washington, DC.

American towns can buy pricey terrorism insurance. The policies do not cover nuclear, biological, or chemical attacks. In order to make a claim, the attack must be certified by the U.S. attorney general, the secretary of state, and the Department of Treasury, and exceed $5 million in damages.

Hartford, Connecticut, became the "Insurance Capital of the World" after the Great Fire of New York in 1835 that wiped out the New York Stock Exchange and most of the country's major insurance companies, who then moved their operations to Hartford.

WHAT THE BLAZES

Every twenty-three seconds fire departments somewhere in the United States are called out to a blaze. That adds up to 1,348,500 calls a year.

Some 219,000 American vehicles go up in flames every year.

Arson is suspected in about 26,500 structural fires each year in the United States.

STAT SHEET

Three in ten Americans say they are reluctant to shake hands for fear of catching germs.

Thirty-three percent of Americans surveyed reported having seen someone extend a hand that they had just coughed or sneezed into.

The average American takes 5,117 steps a day. Sedentary folks take less, while active people may take 10,000 or more steps per day.

There are 450,000 school buses in the United States, which travel about 4.3 billion miles per year. An average of six children a year are killed while riding in school buses.

The average American car gets twenty-one miles per gallon.

The average age of cars in America is ten years.

The Library of Congress adds about fourteen thousand objects to its collection every day.

Since 1991, the U.S. Coast Guard has reported two thousand vessels lost at sea.

The U.S. military has 7 million computers.

Approximately 85 percent of today's U.S. Marines have never been on a ship.

There are eighty-seven thousand flights daily in the United States.

The United States imports 22 percent of its oil from Canada.

Forty percent of the greenhouse gases emitted in the United States come from power plants and oil refineries.

FUNNYTOWN, USA

Seven states are known for their Climaxes—Colorado, Georgia, Michigan, Minnesota, North Carolina, Ohio, and Pennsylvania. There are also scores of other humorous and unusual place-names across the country, including:

Accident, Maryland

Bad Axe, Michigan

Beans Corner Bingo, Maine

Beaverlick, Kentucky

Belcher, Louisiana

Big Lick, North Carolina

Big Ugly, West Virginia

Blue Ball, Ohio

Bobo, Mississippi

Boring, Maryland and Oregon

Bowlegs, Oklahoma

Bugtussle, Tennessee

Bummerville, California

Bumpass, Virginia

Cabbage Patch, California

Camel Hump, Wyoming

Cat Elbow Corner, New York

Cockeysville, Maryland

Colon, Nebraska

Conception, Missouri

Coward, South Carolina

Cumming, Georgia

Deadhorse, Alaska

Defeated, Tennessee

Difficult, Tennessee

Dogtown, California

Embarrass, Minnesota

Fannie, Arkansas

Flippin, Arkansas

Gay and Gay Head, Massachusetts

Goobertown, Arizona

Hell, Michigan

Hell Hollow, Nevada

Hicksville, New York

Hooker, Oklahoma and California

Hoop and Holler, Tennessee

Horneytown, North Carolina

Hot Coffee, Mississippi

Humptulips, Washington

Idiotville, Oregon

Jot 'Em Down, Texas

Jugville, Kentucky

Knockemstiff, Ohio

Licking, Mississippi

Looneyville, Texas

Loveladies, New Jersey

Mutt, Virginia

Nameless, Tennessee

Nimrod, Minnesota

Ninetimes, South Carolina

Nothing, Arizona

Notown, Vermont

Nuttsville, Virgina

Odd, West Virginia

Ogle, Kentucky

Okay, Oklahoma

Oral, South Dakota

Peach Bottom, Virginia

Peculiar, Missouri

Podunk, Michigan

Pointblank, Texas

Possum Trot, Kentucky

Pringle, South Dakota

Ragtown, California

Roachtown, Illinois

Sanatorium, Mississippi

Satans Kingdom, Vermont

Sopchoppy, Florida

Soso, Mississippi

Spread Eagle, Wisconsin

Succasunna, New Jersey

Suck Egg Hollow, Tennessee

Sucker Flat, California

Surprise, New York

Sweet Lips, Tennessee

Threeway, Virginia

Tick Bite, North Carolina

Tightwad, Missouri

Tingle, New Mexico

Toad Hop, Indiana

Toad Suck, Arkansas

Townville, South Carolina

Typo, Kentucky

Ubet, Wisconsin

Uncertain, Tennessee

Waterproof, Louisiana

Weed, California

Worms, Nebraska

Whynot, North Carolina

Why, Arizona

Zap, North Dakota

Zigzag, Oregon

WORLD REPORT

NAME THAT COUNTRY

Côte d'Ivoire is still known as Ivory Coast in many English-speaking countries.

The African nation of Dahomey changed its name to Benin in 1975.

Ghana means "warrior king."

Zimbabwe was formerly known as Rhodesia.

Until 1990, Namibia was known as South-West Africa.

The Democratic Republic of the Congo (not to be confused with the Republic of the Congo, which lies to the west) was formerly known as the Congo Free State, Belgian Congo, Congo-Léopoldville, Congo-Kinshasa, and Zaire.

Belize used to be British Honduras. It is the only Central American country that was a British colony.

The western half of the island of New Guinea is known as West Papua and is part of Indonesia. The eastern half of the island is the country of Papua New Guinea.

Madagascar was formerly known as the Malagasy Republic.

The first person to call Canada by that name was explorer Jacques Cartier.

Bangladesh was known as West Pakistan before gaining independence from Pakistan in 1971.

IN A BIG COUNTRY

Sudan is the biggest country in Africa and the Arab world.

HIGHEST LOW

The country of Lesotho, which lies entirely within South Africa, is the only nation in the world (excluding the Vatican) that is entirely surrounded on all sides by only one country.

Lesotho's low point of 4,593 feet is the highest low point of any country in the world.

THE MIDDLE OF NOWHERE

The least populous and most remote territory in the world is Pitcairn Island, which is inhabited by only about fifty people. Today's population is descended from nine HMS *Bounty* mutineers and the Tahitians they took with them. There are only four family surnames on the island, one being Christian, descendants of the mutiny's leader—Fletcher Christian.

The wreck of the *Bounty*, which the mutineers set ablaze and sank, is still visible on the bottom of Bounty Bay.

COZY COUNTRIES

Gambia is the smallest country on the African mainland and is surrounded by Senegal on three sides with the Atlantic Ocean on the fourth side.

Bhutan is a tiny nation sandwiched between India and China in the Himalayas. In 2006, it was voted by *Business Week* magazine as the "happiest country in Asia."

The island nation of Niue is about 1,500 miles northeast of New Zealand.

Tiny Bangladesh is the seventh most populous country.

Wales is a country of 3 million that is a part of the United Kingdom.

Moldova is a landlocked country between Romania and Ukraine.

Transnistria is a breakaway territory of Moldova that declared its independence in 1990 and has governed itself ever since. No United Nations countries recognize tiny Transnistria.

Abkhazia is a breakaway republic from the country of Georgia. This self-governing area is only formally recognized by Russia, Nicaragua, Venezuela, and Nauru.

Another disputed breakaway republic from Georgia is South Ossetia, which declared its independence in 1990.

The Comoros is an archipelago nation off the coast of Africa between Madagascar and Mozambique.

Tonga is an archipelago nation of 176 islands spread out over 270,000 square miles in the South Pacific.

REVOLVING DOOR GOVERNMENT

In 2010, Japan seated its fifth premier in three years.

YOUNG AND OLD

The country with the largest proportion of its population ages sixty-five and older is Japan, followed by Germany, Italy, Sweden, Greece, and Portugal.

The countries with the largest proportion of its population younger than fifteen are all in Africa. Niger is number one, followed by Uganda, Burkina Faso, the Democratic Republic of the Congo, and Zambia.

CAN'T GET THERE FROM HERE

In 2010, there was a sixty-mile traffic jam in China that lasted ten days.

> The average driving speed in Beijing is fifteen miles per hour.

I SMELL A RAT

In Tanzania, giant rats (the size of small dogs) are used to smell the saliva samples of people to detect tuberculosis. The rats can analyze forty samples in seven minutes. It takes a human one day to do forty samples with a microscope.

BREAKING THE ICE

As of 2010, Iceland is the only country that has an openly gay leader.

LOOKS LIKE RAIN

Parts of Indonesia have three hundred thunderstorms a year.

> The most rain ever recorded in one day—73.6 inches—fell on the isle of Cilaos La Reunion, off the coast of Madagascar in 1952.

TWIN TOWN

The village of Kodinhi, India, is known as "twin town." There have been 220 sets of twins born to just two thousand families. No one knows why.

PORNISTAN

Pakistan prides itself on being "The Land of the Pure." According to the folks at Google, Pakistan leads the world in the number of online searches per person using the keywords "camel sex," "dog sex," "donkey sex," "horse sex," and "rape sex."

SPARE THE ROD

Spanking children is currently illegal in twenty-five countries, including Spain, Portugal, New Zealand, and all of Scandinavia.

On the other end of the disciplining spectrum, in 2010, the highest court in the United Arab Emirates ruled that men can beat their wives and children, as long as they don't leave any bruises, if admonishment and the withholding of sex from the wife doesn't make them submit to the husband's will.

BREATHE YOU IN

The Eskimos, or Inuits, don't rub noses when kissing, but practice a greeting known as *kunik*, where the nose and upper lip are pressed against the cheek or forehead of another and their scent is breathed in.

TOP OF THE WORLD

The Andes are the longest mountain range in the world.

The Andes are still growing and may one day surpass the Himalayas in height.

The largest ice field outside of the polar regions is in the Andes.

GOOD WORK, IF YOU CAN GET IT

Germany and Sweden both guarantee forty-seven weeks a year of paid parental leave. Norway offers forty-four weeks. The United States and Australia offer none.

BIG NEWS

The Great Sphinx of Giza is the largest freestanding stone sculpture in the world.

The biggest clock in the world is located in Mecca, Saudi Arabia. The 130-foot-diameter, four-sided clock sits atop a new 1,970-foot-tall skyscraper.

The tallest building in the world is Burj Khalifa in Dubai. It is 2,716.5 feet tall, or 160 stories high. When completed in 2009, it also became the tallest structure, besting previous record holder the KVLY-TV mast in North Dakota (2,063 feet) and the tallest freestanding structure, topping Toronto's CN Tower (1,815 feet).

The longest rail tunnel in the world is the thirty-four-mile-long Gotthard Base Tunnel, which runs under the Alps and took fourteen years to complete.

TOTALLY AUSS-OME

Ayers Rock in Australia is also called by its Aboriginal name—Uluru.

The Great Barrier Reef is the size of California.

DISTURBING DATA

As of 2010, more than thirty thousand people have been killed in Mexico's drug war. That's many times more than all the Americans who have died in the wars in Afghanistan and Iraq.

A study conducted by the Medical Research Fund found that one in three South African men admit to having raped a woman and one in four South African women say they have been raped.

Rape wasn't a crime in Haiti until 2005.

RENT-A-WIFE

Shi'a law permits men to enter into "temporary marriages" with women and young girls, lasting from a few hours to years. The goal of such "marriages" is to allow the men to have extramarital affairs or visit prostitutes and remain within the teachings of the Koran. Such arrangements are commonplace in Iran, Iraq, Mali, and other Muslim nations and often involve payment to the woman.

BIRDS OF A FEATHER

In 2010, two gay male vultures that had been cohabitating at a zoo in Germany were separated and forced to mate with females. Gay rights activists protested.

SUPER TRAIN

A new train that will travel between Beijing and Shanghai can travel at speeds of 302 miles per hour.

ROYAL FLUSH

Twenty-two countries still have royal families.

STRANGE WORLD

Graham Barker, an Australian librarian, holds the Guinness World Record for the largest collection of his own belly button fluff. Every day since 1984, he has harvested the lint from his navel, about twenty-two grams so far, amassing five large jars of the stuff.

Tatiana and Krista Hogan are British Columbian twins conjoined at the head who see out of each other's eyes.

L. Ron Hubbard, the founder of Scientology, invented a device he called the Hubbard Electrometer, in 1968, to test whether vegetables feel pain. It had metal probes on wires connected to a meter. He used his gizmo to "prove" that tomatoes "scream when sliced."

U.S. military interrogators used Metallica's "Enter Sandman" and Barney's "I Love You" theme song to break Iraqi prisoners.

Robert Keller of Pennsylvania drove his car into the lobby of a Pennsylvania Driver and Vehicle Services lobby, injuring four people, while parking his car after just having passed his road test with the state examiner still in the vehicle.

One Michael Ireland won a $650,000 judgment, in 2010, for an eye injury he received from a lap dancer's high heel at a Florida strip club.

A drunken woman in Owensboro, Kentucky, was charged with assault after she squirted her breast milk in the face of a female jailer while putting on her prison suit. The victim claims the attack was a "biohazard."

Seventy-year-old Indian woman Rajo Devi Lohan became the oldest person in history to give birth, in 2008. She required in vitro fertilization to conceive the child. Unfortunately, her aged body was not up to the delivery. Her womb ruptured and she needed a C-section from which she never recovered, dying in 2010.

In 2010, Mexican matador Christian Hernandez ran out of the ring and jumped over the wall as a bull charged. Hernandez, who had previously been badly gored by a bull, was fined for breach of contract.

> In 2010, a six-month-old infant was killed by a falling tree branch in New York's Central Park. The baby's mother, who was holding her at the time, was injured. Earlier in the year, a forty-six-year-old man was also killed in the park by a falling snow-covered branch. In 2009, a man was knocked out by yet another falling tree branch in Central Park.

In 2010, an eighteen-month-old child in Paris fell out of a seventh-story window, bounced off a ground-floor awning, and was caught by an alert doctor, à la Swee'Pea and Popeye. The toddler was fine.

> Between 2005 and 2010, three different children had their fingers severed by the escalators at Macy's flagship store at Herald Square in Manhattan.

In March 2010, a nineteen-year-old Pennsylvanian was killed while pumping gas when static electricity ignited the fuel. Ten years earlier, an unfortunate Oklahoman suffered the same fate. (As of 2010, there had never been a confirmed report of a cell phone igniting a fire at a gas pump.)

BUY THE LIGHT OF THE MOON

The Outer Space Treaty of 1967, which has been signed by one hundred countries, outlaws nations from owning celestial bodies. This, however, has not stopped some presumptuous people from making individual claims of ownership.

> One Spanish woman, Angeles Duran, claims to own the sun. Her assertion has not been recognized by any governing body.

An entrepreneurial American named Dennis Hope claims to own the moon and actually is selling lots. He sells about 1,500 parcels a day for $20 and has made $9 million doing so. The Canadian government jailed a man for fraud when he tried a similar scheme.

CELLAR DWELLERS

Montreal's Underground City, also known as RÉSO, is the largest underground complex in the world. Its two thousand stores, twelve hundred offices, sixteen hundred housing units, two hundred restaurants, and forty banks cover an area of 1.4 square miles, and it has ten metro train stops.

AN EYE FOR AN EYE

Islamic code in Iran, Pakistan, and Saudi Arabia allows for retribution in violent crimes called *qisas*, where the victim can request that the assailant be punished in a way similar to the crime committed. Thus, when an Iranian woman was blinded in an acid attack in 2009, she demanded her attacker be blinded with acid.

Three of the four schools of Islamic jurisprudence prohibit capital punishment for Muslims who murder non-Muslims. The payment of blood money is allowed. Depending on the Muslim sect involved, the worth of the life of a Christian or Jew is only one-third that of a Muslim and the value of a Zoroastrian is one-fifteenth.

GLOWING REVIEWS

The Ukraine is constructing a massive twenty-thousand-ton shell measuring 345 feet tall, 853 feet wide, and 490 feet long that they are going to slide on rails to cover Chernobyl reactor No. 4, which melted down in 1986. The sarcophagus now covering it is crumbling.

In 1986, the Soviets built the city of Slavutych, fifty kilometers from Chernobyl, to relocate its population. Today, about 255,000 people live there.

The Ukraine plans to open the Chernobyl exclusion zone to tourists in 2011.

NUKE 'EM

The nation most dependent on nuclear energy is Lithuania. They get 76.2 percent of their electricity from nuclear reactors.

CHAMPAGNE WISHES AND CAVIAR DREAMS

The world's most expensive champagne is offered by the Moscow Ritz-Carlton at $275,000 a bottle. Two hundred bottles of the 1907 Heidsieck bubbly sat in a shipwreck on the bottom of the ocean off the coast of Finland for eighty years and was perfectly preserved.

The world's priciest pizza can be found at Manhattan's Nino's Bellissima. The small pie is covered with sliced lobster tail, crème fraîche, and four kinds of caviar. It goes for one thousand dollars. That's about thirty-three dollars a bite.

In 2008, a Burger King in London offered a $190 hamburger, making it the world's most expensive. Simply called "The Burger," it consisted of Japanese Wagyu beef, Spanish cured ham, white truffles, balsamic vinegar, onions, lamb's lettuce, white-wine-and-shallot-infused mayonnaise, and pink Himalayan rock salt, inside a saffron-and-white-truffle-dusted bun. Proceeds for this decadent meat patty sandwich went to a local children's charity.

White truffles from the Piedmont region of France routinely sell for up to $2,700 per pound. The record price paid was $330,000 for one particularly fine three-pound specimen in 2007.

Kopi Luwak is an expensive coffee obtained from beans collected from the droppings of the Asian palm civet, a catlike creature that eats ripened coffee beans. The taste of the beans is apparently improved by their passage through the animal's digestive tract.

'SHROOMS

Mushrooms are the only nonanimal food source that provides vitamin D.

The *Agaricus campestris* mushroom is the one most commonly found in supermarkets. They are sold as white mushrooms or button mushrooms when they are small and unopened. At maturity, when the caps have fully opened and the surface has turned brown, these mushrooms are sold as portobello.

Creekside Mushrooms, in Pennsylvania, the nation's largest mushroom farm, has 150 miles of abandoned limestone tunnels that it uses for cultivation.

Scientists have found a twenty-foot-tall fossilized mushroom that grew 350 million years ago in Saudi Arabia.

EXTREME CUISINE

The locust is the only invertebrate that is considered kosher food.

In Ecuador a favored dish is barbecued guinea pigs.

In the seventeenth century, the Roman Catholic Church, at the request of the bishop of Quebec, ruled that beaver was to be classified as a fish for the dietary restrictions during Lent. The rationale was based on the fact that a beaver is an aquatic animal.

Moose nose was an early American delicacy. Jellied moose nose is still enjoyed by some folks today.

Boca Tacos y Tequila in Tucson, Arizona, offers weekly taco specials featuring such exotic meats as turtle, rattlesnake, elk, alligator, python, kangaroo, and lion.

SWEET SPOT

The average American eats sixty-one pounds of refined sugar every year, including two pounds from Halloween candy.

Eating too much sugar can cause the skin to wrinkle.

More of the sugar consumed in America comes from sugar beets than sugarcane. The sugar from a sugar beet is identical to that from sugarcane.

Aspartame and saccharin were both discovered accidentally when scientists working on research unrelated to sweetness somehow tasted their work.

A 20-ounce Pepsi has 17½ teaspoons of added sugar. A 20-ounce Coke has 16½ teaspoons, and a 16-ounce Snapple Lemon Tea has 10½ teaspoons.

PIG CITY

Cincinnati was once known as "Porkopolis." The city was the main pork-packing center in the 1800s and numerous herds of pigs could be found on the streets.

The largest pig on record was from a farm in Tennessee. It tipped the scales at 2,552 pounds in 1933.

DIET-BUSTERS

The Cold Stone Creamery PB&C (peanut butter and chocolate) Shake has 2,010 calories.

The Friendly's Ultimate Grilled Cheese Burger—a hamburger with two grilled cheese sandwiches for a bun—has 1,160 calories.

The Outback Steakhouse Chocolate Thunder from Down Under, a dessert made of pecan brownies, vanilla ice cream, chocolate sauce, whipped cream, and chocolate shavings has 1,911 calories.

JUST ADD WATER

MRE stands for meals ready to eat. They are self-contained, self-heating meals used by the U.S. military. Thirty-five million are consumed each year.

MREs contain a small pad of magnesium dust, iron dust, and salt. Adding water to the pad creates an exothermic reaction and a temperature of 100°F that heats the meal.

IT MUST HAVE BEEN SOMETHING YOU ATE

Each year, up to 5,000 Americans die from food poisoning, while 76 million get sick from it.

More than two hundred known diseases are transmitted through food.

The number one food responsible for food poisoning is chicken, followed by beef and leafy vegetables.

Noroviruses, which are spread when restaurant workers don't wash their hands after using the toilet, are the most common food-borne illness-causing agents, responsible for 39 percent of illnesses. Salmonella from animal feces is number two, accounting for 27 percent.

Sixty-three percent of American restaurant workers say they have handled or served food while sick.

One in twenty-five outbreaks of food poisoning in restaurants results from eating tainted guacamole and salsa.

APPLES AND ORANGES

The Granny Smith apple was discovered by Australian Maria Ann Smith growing in her garden in 1868.

One reason California oranges are so much brighter than Florida oranges is because of the cooler nights where they are grown. Also, due to California's drier climate, their oranges have thicker skins and are less juicy than Florida oranges.

Brazil leads the world in orange juice production, followed by the United States.

Maraschino cherries are grown in Italy and start out yellow. They are bleached to pure white and then dyed red.

EAT YOUR VEGGIES

Some people lack the ability to metabolize betacyanin, the pigment in beets, and as such, pee bright red after eating them. This phenomenon, which occurs in 10 to 15 percent of people, is known as beeturia.

An artichoke is the flower bud of a plant in the thistle family.

Seventeenth-century cookbooks recommended boiling spinach for twenty-five minutes to remove the toxins that people of the time thought the vegetable possessed.

Iceberg lettuce got its name from the fact that in the 1920s it was shipped from California packed in ice.

Beans give people gas because of their high fiber content and the complex sugar (raffinose) they contain. Humans do not have the enzymes needed to break down raffinose, so bacteria in the lower intestine ferment it, producing foul-smelling gases in the process.

CHIPS AHOY

Potatoes used for chips must be round and have a low moisture content.

Big potatoes are used for chips in big bags, small ones for chips in small bags.

Potato chip bags are flushed with nitrogen gas before filling to extend product freshness.

Pringles are made from potato paste that is rolled out, cut to shape, deep-fried, and salted.

BREAD AND WATER

One bushel of wheat yields about sixty pounds of whole wheat flour, from which forty-two commercial loaves of bread can be baked.

Poland Spring water comes from Poland, Maine, and springs in other Maine locations.

THAT'S NUTS!

Michelangelo used walnut oil as a drying agent in the paint he used on the Sistine Chapel.

Macadamia trees were originally considered ornamental plants, before they were grown for their nuts.

BUSY BEES

It takes 2 million bee trips to flowers to collect enough nectar to make one pound of honey.

EMPTY CALORIES

Cotton candy is known as candy floss in Britain and fairy floss in Australia. It was first brought to the general public's attention at the 1904 World's Fair in St. Louis.

The first corn dog was probably sold at the 1942 Texas State Fair.

Philadelphians eat twelve times the amount of hard pretzels compared to the national average.

There are seventeen Tootsie Roll flavors.

The first Pez dispensers had no heads.

Jellybeans are coated with shellac to make them shiny.

All jellybeans are the same inside. It's the shell that is flavored and colored.

When Kellogg's came out with Sugar Frosted Flakes in 1952, Tony the Tiger had to share the box with his friend Katy the Kangaroo.

NAME THAT FOOD

Russian dressing is an American creation of the turn of the last century. The name comes from the fact that early versions contained caviar.

Catherine de' Medici, who was from Florence, Italy, was extremely fond of spinach, which is why dishes with spinach are known as "Florentine."

Canola is a type of rapeseed developed by Canadian plant breeders in the early 1970s that has a lower erucic acid (monosaturated omega-9 fatty acid) content. The seeds are used to make canola oil. The name was created from "Canadian oil, low acid," which sounded better for marketing purposes than rapeseed oil.

The word "lasagna" comes from the Roman word for "cooking pot," *lasanum*, which comes from the Greek *lasana*, or "chamber pot."

Roman restaurateur Alfredo di Lelio invented fettuccine Alfredo in 1914.

"Canadian" bacon is called "back" bacon in Canada.

THAT'S THE SPIRIT

Up until 1995, it was illegal for beer manufacturers to list the alcohol content on their products sold in the United States. This was because after Prohibition, the government feared that beer makers would engage in strength wars, flaunting their beer's potency to get a larger market share.

Many beers, such as Budweiser and Michelob, still don't note the alcohol content on their labeling.

The beer with the lowest alcohol content is Anchor Small Beer at 3.0 percent, while Samuel Adams Utopia maxes out with a whopping 25 percent alcohol content.

Oktoberfest, held in Munich for sixteen days every fall, is the world's largest fair, with some 6 million visitors annually.

Oktoberfest began in 1810, as a festival to honor the wedding of Crown Prince Ludwig to Princess Therese of Saxe-Hildburghausen.

The daiquiri is named for the beach of the same name near Santiago, Cuba, where a group of American mining engineers invented the drink circa 1900.

MOO JUICE

Milk wasn't first sold in bottles until 1878, in Brooklyn. Before this, city dwellers got their milk from the back of a wagon where it was ladled from barrels into pitchers brought by the buyer.

Finland is the highest per capita consumer of milk, followed by Sweden, Ireland, the Netherlands, and Norway.

Grade B milk, which does not meet government standards to be sold in fluid form, is used for cheese making.

There are moose dairies in Sweden and Russia.

Cow milk contains more protein, but less fat and sugar, than human milk.

A milked cow will give milk two or three times a day for about ten months.

Calves only get mother's milk for a few days and then are put on soymilk formula.

An average cow will produce about 1,400 gallons of milk a year.

Refrigerated milk is still good for five to seven days after the "sell by" date.

The reason skim milk may appear greenish is because the white-imparting fats are removed, while the green-colored riboflavin remains.

Low-fat milk began to outsell whole milk in 1988.

Whole milk has 4 percent fat content, so 1 percent milk has 75 percent less fat than whole milk.

The pasteurization of milk in the United States wasn't required until the late 1920s.

The banana split was first created, in 1904, by David Strickler, who worked in a Latrobe, Pennsylvania, pharmacy.

Rocky Road ice cream was invented by William Dreyer (of Dreyer's Ice Cream fame) in 1929. He picked the name to symbolize the rough times the country was facing during the Great Depression.

Ben Cohen of Ben & Jerry's can't taste very well, which is one reason they add textured mix-ins to some of their ice creams.

Creamsicles are made by putting softened orange-flavored ice cream in a frozen mold. Once the orange ice cream in contact with the mold has frozen, the still liquid center is sucked out and filled with vanilla ice cream.

Monterey Jack cheese is named for dairyman David Jacks, who mass-marketed the cheese first made by the Franciscan friars of Monterey, California.

Mozzarella is America's favorite cheese, thanks to the popularity of pizza.

QUEASY COWS

One-half of the baking soda sold in the United States is to ease the indigestion of cows. Their stomachs are designed to digest grass, not the grains they are fed, so they get gassy.

MEAT MARKET

Store-bought meat is bright red on the outside, but purplish on the inside. This is because oxygen in the air turns the purple-red myoglobin at the meat's surface to oxymyoglobin, which is red. Supermarkets wrap their meats in air-permeable membranes so that they will look nice and red on the shelves.

Beefeaters, the ceremonial guards at the Tower of London, probably got their name because they used to be paid with rations of beef.

In the 1940s, chicken cost more than veal.

FISHY FACTS

Yellowfin tuna began being sold as "fancy" tuna in 1926 to compete with the more expensive albacore.

Light tuna is primarily skipjack.

Americans prefer their tuna packed in water. Europeans like it in oil.

Canned tuna is cooked right in the can.

Half of the fish consumed globally is now farm-raised.

🌰 HOLY MACKEREL

The record price paid for a fish was broken in 2011 when a giant 754-pound bluefin tuna sold for $396,000 in Tokyo. That's $525 per pound.

In the early 1900s, bluefin was known as "horse mackerel" and it was ground into cat food.

Sport fishermen used to pay to have their "worthless" bluefins carted away.

POP CULTURE

The first diet cola was Diet Rite, which came out in 1958.

Phosphoric acid is added to colas to give them a tangy taste and add some zip.

Plastic soda bottles have expiration dates because the PET (polyethylene terephthalate) bottles they come in are permeable to carbon dioxide gas and they will slowly start to go flat over time.

At Passover time, Coca-Cola sells kosher Coke. Instead of high-fructose corn syrup (which is made through fermentation and is not kosher for all Jews at Passover), it is made with sucrose (real sugar). Kosher Coke has yellow caps with Hebrew lettering on the two-liter bottles.

TALKIN' TURKEY

Turkeys are bred with white feathers since the brown feathers of wild birds leave dark spots on the skin of a plucked turkey that consumers would find unacceptable.

Turkeys are bred with such big breasts that mating becomes impossible and the birds must be artificially inseminated.

In the United States, 30 percent of turkeys are eaten for Thanksgiving dinner.

Americans eat the most turkey of any country—seventeen pounds per person per year.

MAMMA MIA

Italy leads the world in per capita pasta consumption at fifty-seven pounds. Venezuela comes in second at twenty-seven pounds. Tunisia, Greece, Switzerland, and the United States follow close behind.

CHOP SUEY

"Chop suey" roughly translates to "odds and ends" in English.

Seventy-five percent of restaurants are fast-food joints.

Paper napkins and paper towels debuted in the 1930s.

In 2010, one in every eight Americans (more than 40 million people) used the USDA's Supplemental Nutrition Program, better known as food stamps.

Heinz ketchup comes in seventeen different kinds and sizes of bottles.

Early cookbooks didn't mention exact quantities because people didn't own measuring cups or spoons.

One tablespoon of soy sauce has 38 percent of the recommended daily allowance of salt.

Anheuser-Busch is the number one user of rice in the United States.

Three billion people in the world eat rice three times a day.

ARE YOU READY FOR SOME FOOTBALL?

THE HAVES AND HAVE-NOTS

The most valuable NFL teams as of 2010 were the Dallas Cowboys at $1.8 billion, followed by the Washington Redskins at $1.55 billion, the New England Patriots at $1.37 billion, the New York Giants at $1.18 billion, and the Houston Texans at $1.17 billion.

The least valuable NFL teams are the Jacksonville Jaguars, the Buffalo Bills, the St. Louis Rams, the Minnesota Vikings, and the Oakland Raiders.

OLD PROS

The first paid football player was former Yale guard William (Pudge) Heffelfinger, who received five hundred dollars from the Allegheny Athletic Association to play in a game against the Pittsburgh Athletic Club in 1892.

The first football team to be composed of entirely professional players was the 1897 Latrobe (Pennsylvania) Athletic Association team.

ATTABOY, TROY

Ex–NFL quarterback Troy Aikman was born with a clubfoot. He was in plaster casts till he was eight months old and wore special shoes until he was three.

Aikman was offered a professional baseball contract by the New York Mets when he graduated high school, but decided to go to the University of Oklahoma to play football.

After being injured in his first season at Oklahoma and losing the starting quarterback job, he transferred to UCLA, where he went on to stardom.

Aikman's ninety wins in the 1990s make him the winningest quarterback of any decade.

CLIMBING THE LADDER

Oakland Raiders owner Al Davis is the only person in pro football history to serve as a personnel assistant, scout, assistant coach, head coach, general manager, commissioner, and team owner. He was once the commissioner of the American Football League (AFL).

WINNING WITH WEEB

Weeb Ewbank is the only coach to have won championships in both the AFL and the NFL. He led the Baltimore Colts to the NFL title in 1958 and 1959, and the
New York Jets to the AFL title in 1968 and the NFL title
in 1969.

SACK MASTERS

Defensive end Deacon Jones first coined the word "sack"
in reference to tackling the quarterback behind the line of
scrimmage during his playing days with the Los Angeles
Rams in the 1960s.

There were no official statistics kept for sacks when
Jones played, but it is believed he recorded 179½ in
his career, which would place him third on the all-
time list.

New York Giants linebacker Lawrence Taylor was known
to his family as "Lonnie."

Taylor is the only defensive player to be unanimously
elected league MVP, in 1986. Minnesota Vikings defensive tackle Alan Page is the only other defensive
player named NFL Most Valuable Player, in 1971.

LIONS AND TIGERS AND BADGERS, OH MY

The following is a look back at some of the long-forgot-
ten teams that once played in the NFL:

The Oorang Indians of LaRue, Ohio, were an all Indian team featuring Jim Thorpe that played in the 1923 and 1924 seasons. They were sponsored by a dog kennel in LaRue and hold the distinction of being the NFL team from the smallest hometown.

The Kenosha (Wisconsin) Maroons played one season in the NFL—1924—and won no games.

The New York Yankees played in the NFL from 1927 to 1928 at Yankee Stadium and featured Red Grange.

The Boston Yanks played from 1944 to 1948, becoming the New York Bulldogs in 1949 and the New York Yanks from 1950 to 1951.

The Boston Patriots became the New England Patriots in 1971.

The Duluth Eskimos began as the Kelley Duluths (after the Kelley-Duluth Hardware Store) from 1923 to 1925, then as the Eskimos from 1926 to 1927.

The Providence (Rhode Island) Steamroller played in the NFL from 1925 to 1931 and won the championship in 1928.

The Akron Pros were a charter member of the NFL in 1920. They changed their name to the Indians in 1926 and folded in 1927.

The Pittsburgh Pirates later became the Steelers.

The Canton Bulldogs played NFL football from 1921–23 and 1925–26. They won the league title in 1922 and 1923 and still hold the NFL record for consecutive games without a loss—twenty-five (including three ties).

The Columbus (Ohio) Panhandles were a charter member of the NFL and played in the league's first game, in 1920, against the Dayton Triangles. They changed their name to the Tigers in 1923 and folded in 1926.

The Dayton Triangles played in the NFL from 1920 to 1929. Their home field was Triangle Park.

The Chicago Tigers played one season, 1920, and were the first NFL team to go under.

Detroit had four early NFL teams. The Heralds were Detroit's first NFL team in 1920. They morphed into the Detroit Tigers in 1921. Later the Detroit Panthers played from 1925 to 1926, and the Wolverines in 1928.

The Orange Tornadoes played the 1929 season in Orange, New Jersey, before moving to Newark in 1930. The team folded after the 1930 season and were reincarnated as the Boston Braves, who would later become the Washington Redskins in 1933.

The Portsmouth (Ohio) Spartans became the Detroit Lions in 1934.

The Hammond (Indiana) Pros played from 1920 to 1926 as a traveling team, not playing any games in their "home" town. Most of their games were played at Chicago's Cub Park (known today as Wrigley Field).

The Rochester (New York) Jeffersons played from 1920 to 1925. Their home games were played at their field on Rochester's Jefferson Avenue.

The Buffalo All-Americans played from 1920 to 1923, then as the Buffalo Bisons from 1924–25, 1927, and 1929, and as the Rangers in 1926. They folded in 1929.

The Staten Island Stapletons were from the Stapleton section of Staten Island and played from 1929 to 1932. In 1929, they fielded the shortest player ever to appear in an NFL game—blocking back Jack "Soapy" Shapiro, who was just five feet, one-half inch tall.

The Cincinnati Reds played a season and a half in 1933 and 1934. They were kicked out of the league in 1934 after eight games for not paying league dues. The St. Louis Gunners played out the rest of their schedule.

The Racine (Wisconsin) Legion played in the NFL from 1922 to 1924. They were inactive in 1925 and played as the Tornadoes in 1926, before going belly-up.

The Milwaukee Badgers played from 1922 to 1926. They went bankrupt in 1926, after the NFL fined them $500 for using four high school players.

The Frankford Yellowjackets played from 1924 to 1931 and won the NFL championship in 1926. They were from Frankford, a section in the northeastern part of Philadelphia.

The Kansas City Blues played in the league in 1924 and played under the name "Cowboys" in 1925 and 1926.

The Tonawanda Kardex, also known as the Lumbermen, were from a suburb of Buffalo, New York. They hold the distinction of having been the team to play in the fewest NFL games, just one in 1921.

The Los Angeles Buccaneers played one season as a traveling team with no home field. They played no games in L.A., but were based out of Chicago and fielded a team of mostly California college players.

The Phoenix Cardinals began in 1898 as the Morgan Athletic Club, on Chicago's south side. The club then became the Normals, the Racine Cardinals (named for Racine Street in Chicago), the Chicago Cardinals, and then the St. Louis Cardinals, before finally moving to Phoenix.

Other old NFL teams that are now extinct include the Brooklyn Lions, the Boston Bulldogs, the Brooklyn Dodgers, the Brooklyn Lions, the Cincinnati Celts, the Cleveland Bulldogs, the Cleveland Rams, the Cleveland Tigers, the Evansville (Indiana) Crimson Giants, the Hartford Blues, the Louisville Breckenridges, the Louisville Colonels, the Minneapolis Marines, the Minneapolis Red Jackets, the Muncie (Indiana) Flyers, the New York Brickley Giants, the Pottsville (Pennsylvania) Maroons, the Rock Island (Illinois) Independents, the St. Louis All-Stars, the Toledo Maroons, and the Washington Senators.

In 1944, the Chicago Cardinals and Pittsburgh Steelers merged for one year, becoming Card-Pitt.

DRAFTY DATA

The first NFL draft of college players was in 1936.

The Philadelphia Eagles selected University of Chicago halfback Jay Berwanger with the first ever draft pick. They traded his rights to the Chicago Bears, but Berwanger never played in an NFL game.

BEST OF THE BEST

Peyton Manning has the highest ever single season passer rating—121.1.

Chad Pennington has the highest career pass completion percentage—66.06.

David Garrard has the lowest ever career interception percentage—2.04.

Oakland Raider punter Leo Araguz holds the record for most punts in a game—sixteen—set in 1998.

Steve O'Neal of the New York Jets holds the record for longest NFL punt—98 yards—set in 1969.

John Madden owns the best regular season winning percentage of any NFL coach with at least one hundred wins. He went 112–39–7 overall.

NEVERS AGAIN

Fullback Ernie Nevers holds the NFL's longest-standing record—he scored forty points in one game while playing with the Chicago Cardinals in 1929. He scored six touchdowns and kicked four extra points.

BREAKING THE BARRIER

The first black professional football player was halfback Charlie Follis, who played for the Shelby (Ohio) AC in 1904.

Fritz Pollard was the first black head coach in the NFL, coaching Akron in 1921.

In 1965, Field Judge Burt Toler became the first black official in NFL history.

GAYS AND DYKES

There are three Gays in the NFL. There are also players with the last names Ah You, Angerer, Belcher, Booty, Breaston, Bushrod, Chick, Colon, Dykes, Fokou, Incognito, Love, Polite, Shirley, Sicko, Tatupu, Whimper, Woody, and Zinger.

CAN YOU SPELL THAT?

During the 2010–11 season, quarterback Ben Roethlisberger and receiver T. J. Houshmandzadeh shared the distinction of having the longest last names in the NFL.

FOOTBALL FACTORIES

The states that produced the most players active in the NFL in 2010 were California with 211, Texas with 181, Florida with 177, Ohio with 85, and Georgia with 80.

> The city that has produced the most NFL players is Houston with twenty-four. Miami is second, with twenty-two, followed by Los Angeles with twenty, Detroit with fifteen, and New Orleans with fifteen.

The college that has produced the most NFL players all-time is Notre Dame with 503. The University of Southern California has produced 437, Ohio State 377, Michigan 331, Penn State 324, Nebraska 311, Oklahoma 296, Tennessee 290, Pittsburgh 278, and LSU and UCLA 272.

> The college football team that had the most future NFL Hall of Famers on its roster was the 1951 San Francisco Dons, who had three—running back Ollie Matson, offensive tackle Bob St. Clair, and defensive end Gino Marchetti.

BOWL 'EM OVER

Alabama has played in the most college football bowl games—fifty-seven. Texas has played in forty-nine bowl games, Tennessee and Southern Cal have been in forty-eight, and Nebraska forty-six.

THE WANDERERS

Over the years, the New York Football Giants played home games at the Polo Grounds, Yankee Stadium, Yale Stadium, and Shea Stadium, before finally settling into Giants Stadium in 1976. They now play in the New Meadowlands in East Rutherford, New Jersey, a stadium that they share with the New York Jets.

SNOW WAY

The Minnesota Metrodome stadium's roof has collapsed four times in its twenty-nine-year history because of snow accumulation. Apparently the roof was not adequately designed to support snow. (Minneapolis averages forty-five inches of the white stuff a year.)

FRIDAY NIGHT LIGHTS

In Texas, in 1930, Spur High School beat Lorenzo High 186–0.

In another Texas high school football record, the Fightin' Indians of Jacksonville eventually defeated the Nacogdoches Golden Dragons after twelve overtimes in 2010. The game lasted five and a half hours.

AMERICA'S OWNER

Dallas Cowboys owner Jerry Jones was born Jerral Wayne Jones, in Los Angeles, in 1942.

Jones made his fortune through oil and gas exploration in Arkansas.

Jones was an offensive lineman and cocaptain of the University of Arkansas football team that won the national title in 1964.

Jimmy Johnson, who Jones hired to coach the Cowboys when he bought them in 1989, was a fellow teammate on that Arkansas team. Another former teammate, Barry Switzer, was also hired to coach the Cowboys in 1994.

Johnson and Switzer are the only coaches to have won both a Super Bowl and an NCAA football championship, Johnson with Miami in 1987 and Switzer with Oklahoma in 1974, 1975, and 1985.

Switzer's father was arrested for bootlegging when he was a boy and he was raised by his mother, who shot herself when he was twenty-one.

THE BIG TUNA

Bill Parcells was born Duane Charles Parcells. He adopted the nickname "Bill" in high school after repeatedly being mistaken for another student with that name.

Parcells was a star quarterback, pitcher, and basketball center for his New Jersey high school teams.

He picked up the nickname "The Big Tuna" while serving as the New England Patriots linebackers coach in 1980.

HOT FOOT

Mark Moseley is the only special teams player to win an NFL MVP award. He got it for his placekicking with the Washington Redskins in 1982.

FLAGGED

Before the institution of penalty flags, football referees used horns and whistles to indicate a penalty.

Penalty flags in football were first thought up by Youngstown State coach Dwight Beede and used in a 1941 game against Oklahoma City University.

The NFL began using penalty flags in 1948.

College football penalty flags were colored red until the 1970s.

NFL penalty flags were colored white until being switched to yellow in 1965.

Canadian Football League penalty flags are colored orange.

Up until 1999, many NFL penalty flags were weighted with metal BBs. During a game in that year, referee Jeff Triplette threw his flag at Cleveland Browns offensive tackle Orlando Brown, hitting him in the eye and doing extensive damage. Brown missed three seasons because of the injury and sued the NFL, winning $25 million in damages. Today, penalty flags are weighted with beans or rice.

BAND AID

The Washington Redskins have an official marching band, as do the Baltimore Ravens.

KANSAS COMET

Gale Sayers, who played running back for the Chicago Bears (1965–71) was known as the "Kansas Comet." He was the youngest person inducted into the Pro Football Hall of Fame, at age thirty-four.

THE GAME WITHIN THE GAME

Fantasy football was the brainchild of some management personnel of the Oakland Raiders and sports writers in the Oakland area, in 1962. They came up with the Greater Oakland Professional Pigskin Prognosticators League (GOPPPL), while the team was on a three-week East Coast road trip, to pass the time.

An estimated 18 million Americans participate in fantasy football.

EXTRA POINTS

In 1925, NFL teams were limited to sixteen-player rosters, with many players playing both offense and defense.

In 1965, football overtook baseball as America's favorite sport, according to a Harris survey.

The NFL adopted the sixteen-game schedule for the 1978 season.

In December 1925, the Chicago Bears went on a barnstorming tour, playing games in eight different cities in just twelve days.

The first NFL game broadcast nationally was the Thanksgiving Day 1934 contest between the Chicago Bears and Detroit Lions that aired on NBC radio.

The New York Giants and the San Francisco 49ers have retired the most jersey numbers—eleven each. The Baltimore Ravens, Houston Texans, Jacksonville Jaguars, Oakland Raiders, and Dallas Cowboys have retired none. The Seattle Seahawks retired the number "12," which is supposed to represent the "twelfth man"—their fans.

The Green Bay–Appleton television market ranks number seventy in the nation, lowest of any NFL team playing area.

The first football game played indoors was the 1902 contest between New York and Syracuse at Madison Square Garden before three thousand spectators.

SPORTS OF ALL SORTS

MARATHON MATCH

The longest professional tennis match ever played was an eleven-hour-and-five-minute marathon between John Isner and Nicolas Mahut at Wimbledon in 2010.

The final set lasted a record eight hours and eleven minutes, which was almost two hours longer than the previous longest set.

The 183 games played stretched over the course of three days. The final set was 138 games.

Both players held service for a record 168 consecutive games.

Isner finally won.

ACES

Rod Laver has the all-time best winning percentage in men's tennis major finals—.833—with a 5–1 win/loss record. He is followed by Rafael Nadal, Pete Sampras, Roger Federer, and John Newcombe, in that order.

KEEP YOUR EYE ON THE BALL

There is only one referee in a World Cup soccer match. By comparison, NFL football games have seven officials on the field.

THE FLYING TOMATO

Olympic gold medalist snowboarder Shaun White has his own private half-pipe built into the side of a mountain in Silverton, Colorado.

Shaun White is the first person to compete in and win gold medals in both the Summer X Games (where he won in skateboarding) and the Winter X Games (where he won in snowboarding).

RIGHT ON THE MONEY

The highest paid coaches in professional sports, as of 2010, are Phil Jackson of the Los Angeles Lakers, with an annual salary of $10.3 million; Bill Belichick of the New England Patriots with an annual salary of $7.5 million; and Mike Shanahan of the Washington Redskins, Pete Carroll of the Seattle Seahawks, and Larry Brown of the Charlotte Bobcats with an annual salary of $7 million each.

 LOSING PROPOSITION

The NBA lost $400 million during the 2009–10 season.

WHAT'S WRONG WITH THIS PICTURE?

Pro Cycling's Tour of California is sponsored by Amgen—the company that makes the banned endurance performance enhancer EPO.

SPEAKING OF STREAKING

The longest winning streak in Division I college sports belongs to the Miami, Florida, men's tennis team, which had 137 consecutive victories between 1957 and 1964.

The Penn State women's volleyball team had a 109 match winning streak snapped by Stanford in 2010.

The longest winning streak in NCAA Division I football belongs to Oklahoma, who won forty-seven straight games between 1953 and 1957.

The longest unbeaten streak in NCAA Division I football is held by the Washington Huskies, who remained unbeaten for sixty-three consecutive games from 1907 to 1917.

The longest losing streak in NCAA Division I football belongs to Northwestern, who lost thirty-four straight games between 1979 and 1982.

Prairie View A&M University, a Division I-AA school, lost eighty consecutive football games from 1989 to 1998.

In 1899, the University of the South, in Tennessee, won five college football games on the road in six days, beating Texas, Texas A&M, Tulane, LSU, and Ole Miss, by a combined score of 91–0.

The longest win streak in any professional sport was Pakistani Jahangir Khan's 555 consecutive wins in squash, from 1981 to 1986.

The Los Angeles Lakers had the longest NBA winning streak of thirty-three games in the 1971–72 season.

The NBA's Cleveland Cavaliers hold the record for longest losing streak—twenty-four games in 1982.

The Boston Celtics won eight straight NBA championships from 1959 to 1966.

Wilt Chamberlain had seven straight NBA games of scoring fifty or more points in the 1961–62 season.

The NBA consecutive-games-played streak belongs to A. C. Green, with 1,192, while playing with several different teams.

The New York Giants had the longest win streak in Major League Baseball with twenty-six in 1916.

The Atlanta Braves won fourteen straight National League East division titles beginning in 1991.

The Philadelphia Phillies hold the modern major league record for most baseball games lost in a row—twenty-three.

The 1899 Cleveland Spiders professional baseball team holds the record for losses on the road in a season. Because home attendance was so low at their games, many opposing teams refused to play there. As a result, they played 112 away games, losing 101, a record that will never be broken because major league teams today play only 81 road games in a season.

The Spiders lost twenty-four straight games and forty of their last forty-one in 1899. They were so bad that they were relegated to the minor league after the season.

Carl Hubbell of the New York Giants is the pitcher with the longest win streak of twenty-four consecutive games from 1936 to 1937.

Los Angeles Dodgers pitcher Orel Hershiser pitched fifty-nine consecutive scoreless innings between 1988 and 1989.

Willie Mays has played the most Major League Baseball games at centerfield—2,677.

Jim Marshall, who played defensive end for the Cleveland Browns and Minnesota Vikings, holds the NFL record for consecutive games played by a defensive player—282.

The Indianapolis Colts hold the NFL record for consecutive regular season wins with twenty-three from 2008 to 2009.

Receiver Jerry Rice holds the record for consecutive games with at least one catch—274.

The longest winning streak in high school football belongs to De La Salle High School in California, which won 151 games in a row from 1992 to 2004.

The golfer with the most consecutive wins on the PGA Tour is Byron Nelson, who won eleven straight in 1945.

The Philadelphia Flyers remained unbeaten for thirty-five straight games during the 1979–80 season.

The NHL's Calgary Flames hold the record for most consecutive games without being shut out—264—between 1981 and 1985.

Wayne Gretzky of the NHL's Edmonton Oilers had a fifty-one game streak of scoring at least one point, in the 1983–84 season.

Doug Jarvis holds the NHL consecutive games played streak, with 964, while playing for three different teams.

Chris Evert holds the record for consecutive tennis matches won on clay courts at 125.

Martina Navratilova has won the most consecutive tennis matches on any surface, with seventy-four in 1984.

The New Jersey Institute of Technology (NJIT) High-landers hold the NCAA Division I record for most con-secutive basketball losses—fifty-one.

Heavyweight boxer Rocky Marciano, who was un-beaten in his professional career, won forty-nine consecutive bouts between 1947 and 1956.

Middleweight boxer Sugar Ray Robinson had ninety-one straight wins from 1943 to 1951.

IT'S OUTTA HERE

Only two MLB players have ever hit a grand slam on the first pitch they faced in the big leagues. The most recent player was Boston Red Sox rookie Daniel Nava in 2010. Only four players have hit grand slams in their first at bat.

TWENTY BY FOUR

Only twice in the history of Major League Baseball has one team had four twenty-game winners in a season. The Chicago White Sox did it in 1920, with pitchers Red Faber at 23 wins, Lefty Williams at 22, and Dickie Kerr and Eddie Cicotte, both at 21. The 1971 Baltimore Orioles did it with Dave McNally winning 21 and Jim Palmer, Mike Cuellar, and Pat Dobson each winning 20.

WINNER'S CIRCLE

Bobby Cox and Joe Torre are the MLB managers with the most postseason appearances with fifteen each. Tony La Russa has thirteen and Casey Stengel has ten.

THE BOSS

George Steinbrenner and a group of investors, including carmaker John DeLorean, bought the Yankees for $8.8 million in 1973. When George died, in 2010, the team was worth over $1 billion.

In Steinbrenner's first twenty-three years of running the Yanks, he went through twenty managers. He hired and fired Billy Martin five different times.

Steinbrenner was convicted of making illegal campaign contributions to Richard Nixon in 1974 and was suspended from baseball for fifteen months by commissioner Bowie Kuhn.

In 1990, he was suspended from baseball for life (later reduced to two years) for hiring a Mafia affiliated gambler to dig up dirt on star player Dave Winfield, with whom he was having a feud.

THE HOUSE THAT RUTH BUILT

The house where Babe Ruth was born used to stand where centerfield is now in Baltimore's Camden Yards.

Babe Ruth died from throat cancer after years of chewing tobacco.

ICY HOT

Wayne Gretzky is the NHL player who reached one thousand points the fastest—in 424 games. He is followed by Mario Lemieux (513 games), Mike Bossy (656 games), Peter Stastny (682 games), Jan Kurri (716 games), Guy Lafleur (720 games), and Bryan Trottier (726 games).

YOUNG BLADE

The youngest player ever drafted by an NHL team was Garry Monahan, in 1967, by the Montreal Canadians, at age sixteen years, seven months.

STANDING ROOM ONLY

The largest crowd ever for a hockey game was at the 2010 Michigan versus Michigan State college game in Michigan Stadium in Ann Arbor, Michigan. A crowd of 113,411 watched Michigan triumph 5–0.

LUCKY LOGO

The Boston Celtics logo of a leprechaun spinning a basketball on his finger was created by Zang Auerbach, brother of legendary Celtics coach Red Auerbach, in the early 1950s.

FREAKY FOLLIES

Chicago Cub Sammy Sosa had to go on the disabled list when his back went into spasms after sneezing twice just before a 2004 Major League Baseball game against the San Diego Padres. Three other players—Marc Valdes, Russ Springer, and Goose Gossage—have also been forced onto the disabled list after injuring their backs while sneezing.

Boston Red Sox star Wade Boggs once hurt his shoulder when he fell in a hotel room while trying to take off his cowboy boots.

San Diego Padres pitcher Adam Eaton missed a game when he stabbed himself in the stomach with a small knife while trying to get the shrink-wrap off of a DVD.

In 2001, Oakland A's rookie Mario Valdez had his career end when he twisted his ankle stepping off the team bus.

Major Leaguers Brian Jordan and Don Aase both hurt their backs picking up their children.

In 1995, Florida Marlins second baseman Brett Barbarie missed a game after cutting up chili peppers and then putting in his contact lenses, causing his eyes to swell up and burn.

In 1994, Steve Sparks of the Milwaukee Brewers dislocated his shoulder when he attempted to rip a phone book in half.

Detroit Tigers pitcher Joel Zumaya missed the 2006 American League Championship Series because his wrist became sore after playing the video game *Guitar Hero* a little bit too enthusiastically.

Outfielder Marty Cordovan missed several games with the Baltimore Orioles after he fell asleep in a tanning bed and got a nasty burn. His doctor told him to avoid sunlight.

Slugger Kevin Mitchell was placed on the disabled list after pulling rib muscles while vomiting too strenuously.

In 2010, Los Angeles Angels first baseman Kendrys Morales fractured his left leg when he leaped onto home plate while celebrating his game-ending grand slam, ending his season.

In 1997, Washington Redskins quarterback Gus Frerotte injured his neck after head-butting a wall at the back of the end zone while celebrating a touchdown.

NFL placekicker Bill Gramatica tore a knee ligament while celebrating a field goal while with the Arizona Cardinals in 2001. The injury ended his season.

In July 2010, Serena Williams stepped on broken glass at a Munich restaurant, requiring eighteen stitches and ending her playing for the rest of the year.

GAME CHANGER

After the 1967 season, college basketball banned dunking, largely because seven-feet-two-inch Lew Alcindor (who later changed his name to Kareem Abdul-Jabbar) was so dominant. Dunking was permitted again in 1976.

SIMPLER TIMES

In 1964, there were only six teams in the NHL and nine teams in the NBA.

THAT'S AMORE

TALK DIRTY TO ME

Women are more likely than men to talk dirty.

Women with PhDs are more likely than women with BAs to be interested in one-night stands.

SURVEY SAYS . . .

One of the largest sex surveys conducted in the United States found that those surveyed had participated in a total of forty-one different kinds of sex acts in their last sexual encounter.

Eighty-five percent of men reported bringing their female partners to orgasm in their last encounter, while only 64 percent of women reported having one.

One-third of women reported experiencing pain during their last session. Only 6 percent of men did.

Seven percent of women and 8 percent of men report being gay or bisexual.

Fourteen percent of men say they did not enjoy sex their first time, in contrast to 60 percent of women.

DO NOT ENTER

Vaginismus is an uncontrollable tightening reflex in the pubococcygeus muscle that makes any kind of penetration of the vagina very painful or impossible. Causes may be psychological or physical.

IN THE SWIM OF THINGS

Sperm can swim seven inches in an hour.

It takes sperm ten weeks to reach maturity.

Sperm are held ready for action in the epididymis, located on top of the testicles. Sperm may wait there for several weeks. (Women have a vestigial epididymis called a Gartner's duct.)

Sperm compose just 5 percent of semen. The rest is protective and nutritive fluid.

Once sperm enter the uterus, they can survive for up to five days.

Since Y chromosomes are much smaller than Xs, Y-chromosome-bearing sperm weigh less than X-chromosome-bearing sperm and as such are faster swimmers. However, X-chromosome-bearing sperm are more robust and live longer.

New research has found that contact with sperm may increase a woman's chances of ovarian cancer. Studies show that any kind of birth control, even vasectomy, reduces a woman's odds of ovarian cancer by 40 to 65 percent.

SCENT OF A WOMAN

Chemical signals found in women's tears are a turnoff for men. Studies have found that men who sniff these chemicals experience a drop in testosterone levels and become less interested in sex.

Recent studies have found that women appear most attractive to men when they are ovulating.

WAS IT GOOD FOR YOU, TOO?

During an orgasm, the blood flow to a woman's brain basically shuts down. Some women even pass out.

The blood flow to the anxiety center of a man's brain loses blood flow during an orgasm, but the other parts of the brain stay alert.

Twenty-five percent of college-age men surveyed said they have faked an orgasm.

🥜 COME ON NOW

Persistent genital arousal disorder causes sufferers, usually women, to be sexually aroused almost constantly. Orgasms may offer temporary relief. Some of those afflicted may have up to fifty orgasms a day.

PUCKER UP

A romantic kiss causes the eyes to dilate.

When a man exchanges his saliva with a woman during passionate kissing, he also passes along some of his testosterone. Frequent kissing can lead to the buildup of testosterone in the female and increase her libido, making her more open to sexual advances.

FOR BETTER OR WORSE

The median age for first marriages in the United States is 28.1 for men and 25.9 for women. The average age of first-time mothers is 25.

Thirty-one percent of American marriages are interfaith.

A recent study found that when people get married their cardiovascular fitness goes down. Men who got divorced during the study experienced an increase in cardiovascular health, while women did not.

The state with the lowest divorce rate is Massachusetts, which has a rate less than one-third that of Nevada, America's divorce leader.

GO FORTH AND MULTIPLY

One in five American women who are past their child-bearing years are now childless.

The most educated American women are the most likely to be childless.

Between 1 and 2 percent of American women undergo in vitro fertilization. It has about a 50 percent success rate and costs a minimum of twenty-four thousand dollars.

In 2008, 19 percent of births in the United States were to mothers aged thirty-five and older.

OH, BABY

About one in four hundred fraternal twins born to white women in the United States have different fathers. Some estimates put the worldwide number between one in ten and one in fifteen.

A recent Australian study found that boys who are breastfed until at least the age of six months do better on reading, spelling, and math tests at age ten than those who aren't. Curiously, the girls studied showed no difference in performance.

Only 43 percent of American mothers are still breast-feeding their babies six months after giving birth.

BRIEF ENCOUNTERS

The average American woman has twenty-one pairs of panties. Ten percent own thirty-five or more pairs.

Briefs are the most popular panty style.

Fifty-six percent of women fold their panties before putting them away. Twenty-seven percent just throw them in a drawer.

Thirty percent of American women complain of getting wedgies, and 27 percent say ill-fitting or ugly panties can ruin their day.

Ten percent of women report going out sans panties.

SEEDY STATS

Each year, the American porn industry puts out eleven thousand videos.

Eighty percent of legal American porn movies come out of the San Fernando Valley.

Porn actors are required by law to test negative for HIV and STDs thirty days before filming.

The Adult Industry Medical Healthcare Foundation caters to adult film stars.

RED-LIGHT STATE

In 2004, there were roughly 14,500 prostitution-related arrests in California and only three in Vermont.

DON'T CALL US

Among the women Hugh Hefner has vowed will never grace the pages of *Playboy* are Kate Gosselin, Britney Spears, and Kelly Osbourne.

BEHIND THE MUSIC

FIVE-DOLLAR DAD

"Father of Mine" by Everclear was written by lead singer Art Alexakis about his dad. Apparently, his father really did send him Christmas cards containing five dollars.

"Volvo Driving Soccer Mom" is Everclear's song about teen girls who enjoy sex and drugs and grow up to be conservative mothers.

WHAT SONG IS IT YOU WANT TO HEAR?

Lynyrd Skynyrd's record company thought "Free Bird" was too long and didn't want to put it on an album. The band always plays the song as their final encore, after lead singer Johnny Van Zant asks the crowd, "What song is it you want to hear?"

Johnny Van Zant is the younger brother of original Lynyrd Skynyrd singer and founder, Ronnie Van

Zant. His other brother, Donnie Van Zant, founded .38 Special.

Ronnie Van Zant wrote "That Smell" after guitarist Gary Rossington's 1976 car crash—the lyrics contain references to whiskey bottles and an oak tree.

STANDIN' ON THE CORNER

Jackson Browne started writing the song "Take It Easy," but couldn't figure out how to end it, so he gave the tune to his friend Glenn Frey, who had just started his group, the Eagles, and needed material. The Eagles finished the song and it became their first single.

There is a Standin' on the Corner Park in Winslow, Arizona, that commemorates the song.

HARRY, KEEP THE CHANGE

Harry Chapin really was a cab driver in New York, which helped to inspire his song "Taxi."

"Cat's in the Cradle" is based on a poem written by Harry Chapin's wife—Sandy.

RAMBLIN' MEN

The Allman Brothers' "Ramblin' Man" was based on a Hank Williams song of the same name.

Gregg Allman wrote "Whipping Post" on an ironing board in the middle of the night using burned match ends because he could not find a pen.

Gregg Allman had the whole song "Melissa" written except the name Melissa. The name finally came to him in a supermarket when a mother called after her daughter Melissa to come back.

"Jessica" is the name of the daughter of former Allman Brothers guitarist Dickey Betts. She was one year old when he wrote the song. Betts wrote the song "Blue Sky" for Jessica's mother Sandy "Bluesky" Wabegijig, a Native American he married in 1973 and divorced two years later.

TOYS IN THE ATTIC

Steven Tyler left the original lyrics to Aerosmith's "Walk This Way" in the cab he had taken to the recording studio. He rewrote the song on the studio wall. The title came from the movie *Young Frankenstein*, where Igor (Marty Feldman) tells Dr. Frankenstein to "Walk this way," which Gene Wilder's character does to great comedic effect.

Aerosmith's "Janie's Got a Gun" is about a girl who kills her father for sexually abusing her.

"Toys in the Attic" is from an old expression meaning "crazy," similar to "bats in the belfry."

FAB FOUR

Strawberry Field was a Salvation Army children's home in Liverpool that John Lennon would visit on occasion as a teen.

> Only Paul and Ringo were involved on the recording of "Why Don't We Do It in the Road."

Paul wrote the music for "When I'm Sixty-four" at age fifteen. He later wrote the lyrics for his dad's sixty-fourth birthday.

> John Lennon wrote the first two lines to "I Am the Walrus" while on acid. He wanted to write a totally nonsensical song to mess with the heads of all the people who were analyzing Beatles songs and finding meanings that weren't there. Much of it was based on the poem "The Walrus and the Carpenter" from *Through the Looking-Glass*, by Lewis Carroll.

"Rocky Raccoon" was originally titled "Rocky Sassoon."

> A Magical Mystery Tour was a bus trip to an unknown destination popular in England at the time the song of the same name was released.

"Happiness Is a Warm Gun" was the slogan of the National Rifle Association that John Lennon happened upon in a magazine.

George Harrison wrote the music for "Taxman" after the theme song to the hit 1960s TV show *Batman*, of which he was an avid viewer.

"Say, Say, Say," a duet by Paul McCartney and Michael Jackson, is ranked as the best performing Michael Jackson single on the *Billboard* Hot 100.

It took the Beatles seven hundred hours to record the *Sgt. Pepper's Lonely Hearts Club Band* album. Their first album—*Please Please Me*—was recorded in ten hours.

"She Came in Through the Bathroom Window" recounts the time a female fan actually crawled through Paul's bathroom window.

Paul's father wanted him to sing "Yes, yes, yes" instead of "Yeah, yeah, yeah" in "She Loves You" because he thought it sounded more dignified.

"She's Leaving Home" was inspired by a newspaper story Paul had read about a girl—Melanie Coe—who had run away from home. None of the Beatles played instruments on this piece.

Paul McCartney wrote "Let It Be" about a dream he had about his deceased mother, Mary.

In the Wings song "Let Him In," Paul McCartney sings about his aunt (Auntie Gin), his brother (brother Michael), the Everly Brothers (Phil and Don), Keith Moon (Uncle Ernie), and John Lennon (brother John).

Paul McCartney got the title for "Ob La Di, Ob La Da" from reggae band Jimmy Scott and the Obla Di Obla Da Band. Scott tried to get compensation from the Beatles to no avail. However, once when Scott was arrested, Paul McCartney arranged for his bail in return for Scott giving up rights to the name. This song was never released as a single, as the other three Beatles hated the tune.

Ringo Starr wrote "Octopus's Garden" after a boat outing in Sardinia where the captain told him all about the habits of the octopus.

The Beatles' song "Norwegian Wood" is about a cheap pine paneling people once used to decorate their homes.

"Helter Skelter" was the name of an amusement park slide in Britain.

"Penny Lane" was written by Paul about the Penny Lane bus station where he often met John. It was a hub they used to go to other places in London. The shelter referred to in the song is now a restaurant— the Sgt. Peppers Bistro.

The Wings' song "Jet" is about Paul McCartney's Labrador retriever puppy.

"Julia" was written by John Lennon for his mother, who was killed by a car in 1958.

In Japanese, "Yoko" means "ocean child."

BAD SCOOTER SEARCHING FOR HIS GROOVE

Bruce Springsteen got the title "Thunder Road" from a poster for the 1958 Robert Mitchum movie of the same name. He never saw the film.

Bruce's first car was a '57 Chevy with flames painted on the hood.

Bruce Springsteen says he doesn't know what a "Tenth Avenue Freeze-Out" is. Tenth Avenue intersects with E Street in Belmar, New Jersey. The name Bad Scooter in the song refers to Bruce Springteen (same initials).

In "Rosalita (Come Out Tonight)" Springsteen sings about a record company giving him a big advance. He's referring to the twenty-five-thousand-dollar advance he got from Columbia Records.

THE BITCH IS BACK

In "The Bitch Is Back," Elton John is singing about himself. One day when he was complaining a lot around his lyricist Bernie Taupin's wife, she quipped, "Uh-oh, the bitch is back." Taupin liked the line and turned it into a hit song.

The line in Elton John's "Tiny Dancer" that refers to a band seamstress was written by Bernie Taupin about his girlfriend Maxine Feibelman, who traveled with the group and would sew their costumes back together when needed.

"Philadelphia Freedom" was written for tennis star Billy Jean King, who was the coach of the World Tennis Team Philadelphia Freedom in 1974. Elton was a great admirer of hers.

PIANO MAN

In "The Entertainer," Billy Joel's line referring to cutting down a hit to 3:05 is a reference to his record company cutting the length of his hit "Piano Man."

"Piano Man" is Billy Joel's recollections of the six months he spent playing at the Executive Room in L.A.

Before Billy Joel became successful, he was so despondent over losing a girlfriend that he drank a bottle of furniture polish to try to kill himself. He spent three weeks in a mental ward to get his head together.

Billy Joel wrote "She's Always a Woman to Me" about the wife (Elizabeth Small) of the drummer in an early band he had been in that he stole away and married.

Billy Joel wrote "Just the Way You Are" as a birthday present for his first wife Elizabeth.

Billy Joel's "Only the Good Die Young" did poorly when it was first released. Once a New Jersey Catholic university banned the song from its radio station, interest piqued and sales took off. Joel wrote "Only the Good Die Young" about Virginia Callahan, a girl he once had a crush on.

Billy Joel's "New York State of Mind" is from the 1976 album *Turnstiles*. He wrote the song on his way home to New York after spending four years in L.A. Also on *Turnstiles* is "Say Goodbye to Hollywood," which he wrote after his return from California.

Joel wrote "Miami 2017 (Seen the Lights Go Out on Broadway)" about being an old man and moving to Florida and the apocalypse that follows.

BACK IN THE BLACK

The album *Back in Black* by AC/DC has sold 49 million copies and is the second-bestselling album worldwide ever. It is the top-selling album by a band ever.

The title for the AC/DC song "Dirty Deeds Done Dirt Cheap" comes from the villain "Dishonest John" in the old *Beanie and Cecil* cartoons who carried business cards with this motto.

AC/DC guitarist Angus Young got the idea for the title "For Those About to Rock (We Salute You)" from a book about the gladiators in ancient Rome—*For Those About to Die, We Salute You.*

"Hell's Bells" was written as a tribute to lead singer Bon Scott, who drank himself to death a few months earlier. It begins with a bell tolling thirteen times.

"The Jack" by AC/DC is about venereal disease. It was originally titled "The Clap."

SPREAD THE NEWS

Huey Lewis and the News recorded several versions of "The Heart of Rock & Roll," which were sent out to radio stations around the country mentioning the nearest city in order to get airplay.

FOO FACTS

"The Pretender" by the Foo Fighters holds the record for most weeks atop the Modern Rock Tracks chart with seventeen.

"Let It Die" by the Foo Fighters is about how songwriter Dave Grohl felt Courtney Love led Nirvana frontman Kurt Cobain to drugs and ultimately death.

A MOMENT OF SILENCE

The first recording of Simon & Garfunkel's "The Sounds of Silence" was all acoustic. When the song and their first album bombed, selling only two thousand copies, Paul Simon and Art Garfunkel split up. Unbeknownst to them, a producer at Columbia Records overdubbed the song with electric instruments and the song became a huge hit, prompting Simon & Garfunkel to get back together. Later releases of the song are titled "The Sound of Silence."

THRILL SEEKER

The video for Michael Jackson's "Thriller" was voted top music video of all time by VH1.

Michael Jackson had a disclaimer at the beginning of the video stating that he in no way endorsed the occult. He was a Jehovah's Witness.

"Thriller" was originally titled "Starlight."

The UPC price code on the back of the *Thriller* album contained seven numbers that were rumored to be Jackson's phone number. Because of this, thousands of people around America in different area codes received annoying phone calls from fans.

Michael Jackson "borrowed" the "Mama-say, ma-ma-sa, ma-ma-coo-sa" line in "Wanna Be Startin' Somethin'" from the popular 1972 Caribbean song by Manu Dibango—"Funky Soul Makossa." MJ settled with Dibango out of court.

Michael Jackson wrote "Billie Jean" about a woman who was stalking him, claiming that he fathered her child.

YOU'RE THE POET IN MY HEART

"Sara" by Fleetwood Mac is about singer/model Sara Recor, who helped Stevie Nicks write the tune. Nicks was earlier having an affair with the married Mick Fleetwood. Recor would later steal Fleetwood away from Nicks and marry him.

"Rhiannon" is the name of a Welsh goddess. Stevie Nicks wrote the song in ten minutes after reading the novel *Triad*, by Mary Leader, which is about a woman named Charlotte who is possessed by a woman named Rhiannon. Nicks loved the name.

QUIT YOUR DAY JOB

Boston guitarist Tom Scholz worked on "More Than a Feeling" for five years in his basement. It was inspired by the 1967 Left Banke song "Walk Away Renee."

"Rock and Roll Band" by Boston is not autobiographical. Instead of paying their dues playing clubs in Massachusetts, Boston had never toured before their debut album became an overnight sensation.

Scholz worked as a senior product design engineer for Polaroid during the day and wrote and recorded music in his home at night.

YOU BETTER RUN

Pink Floyd's "Run Like Hell" was inspired by Hitler's rise to power and Kristallnacht.

"Shine on You Crazy Diamond" is a tribute to original Pink Floyd member Syd Barrett, who was fired from the band and went crazy. The first letters of Shine, You, and Diamond spell out "Syd."

Before Barrett was kicked out of the band, in 1968, he sometimes refused to perform or would play the same note over and over.

Roger Waters wrote "Vera" about World War II–era singer Vera Lynn, who entertained the troops.

DUTCH TREAT

Drummer Alex Van Halen and his brother, guitarist Eddie Van Halen, were both born in Amsterdam, Netherlands.

Van Halen's "Runnin' with the Devil" opens with the backward blare of car horns. Eddie Van Halen achieved this effect by wiring together several car horns to a car battery and attaching the contraption to a foot switch.

Before David Lee Roth had a hit with "Just a Gigolo," it was a hit for Bing Crosby in 1931.

"Jump" was Van Halen's only number one hit with David Lee Roth.

Van Halen's *5150* album gets its name from the California police code for a mentally unstable person who is to be confined for their own protection.

Van Halen's "Panama" is about a race car called "Panama Express" that David Lee Roth saw in Las Vegas. His own car was also named Panama.

Eddie Van Halen wouldn't take any money for his guitar work on Michael Jackson's "Beat It."

A DAY IN THE PARK

Chicago's main songwriter/keyboardist, Robert Lamm, wrote "Saturday in the Park" after spending one Fourth of July in New York's Central Park watching various performers.

Robert Lamm wrote "25 or 6 to 4" after looking at the time on his watch while trying to come up with a line to finish the song's chorus.

SHE WORKS HARD FOR THE MONEY

Donna Summer wrote "She Works Hard for the Money" about a bathroom attendant she found sleeping in Chasen's restaurant in West Hollywood, California. The woman explained that she was exhausted from working two jobs.

While singing the erotic moaning part of "Love to Love You Baby," Donna Summer lay on the floor and had all the studio lights turned off.

PSYCHO SONGS

Tears for Fears were big fans of psychologist Arthur Janov's Primal Therapy. Their song "Shout" was inspired by the screaming therapy used to get patients to confront their fears, although the song is more about political protest.

The Tears for Fears album title *Songs from the Big Chair* is taken from the movie *Sybil*, where the girl with multiple personalities only feels safe sitting in her psychiatrist's big chair.

THE REAL THING

The U2 song "Even Better Than the Real Thing" is a poke at the Coca-Cola slogan, "It's the Real Thing," and commercialism in general.

U2 used the name Feedback when performing in their high school talent show.

In 1997, the TV special *U2: A Year in Pop* became the lowest rated nonpolitical documentary in the history of the ABC television network.

SPACE CADET

David Bowie wrote "Space Oddity" after seeing the movie *2001: A Space Odyssey*.

"Ashes to Ashes" was written by Bowie in 1980 as a follow-up to "Space Oddity," where Major Tom is still stranded in space, but is happy about it.

BUYING A STAIRWAY

"Stairway to Heaven" has sold more sheet music than any other rock song.

> "Livin' Lovin' Maid (She's Just a Woman)" by Led Zeppelin was about one of the band's groupies. They never played it in concert because guitarist Jimmy Page hated the song. They also never played "Ramble On" live either.

Led Zeppelin's "When the Levee Breaks" is a remake of a tune about the Great Mississippi Flood of 1927. The original was written in 1929.

SUITE MUSIC

Stephen Stills wrote "Suite: Judy Blue Eyes" about his girlfriend, folksinger Judy Collins. He called it "Suite" because he linked together several snippets of prose he had written about her over several months.

> Graham Nash wrote "Our House" about the time he lived with Joni Mitchell in a Laurel Canyon, California, cottage in 1969.

REM-ARKABLE READING

"Orange Crush" by R.E.M. is about Agent Orange, the defoliant sprayed on Vietnam during the war, not the soft drink.

"Losing my religion" is a Southern expression meaning "at wit's end."

C-O-L-A COLA

The BBC made the Kinks change the words in "Lola" from "it tastes just like Coca-Cola" to "it tastes just like cherry cola" because they refused to allow a product name to be mentioned in a song. This is why there are two versions of the hit tune.

FRANK'S WAY

The Frank Sinatra hit "My Way" is an old French song rewritten by Paul Anka.

"New York, New York" was first sung by Liza Minnelli in the movie of the same name in 1977.

THE LONG AND SHORT OF IT

The Guns N' Roses hit "November Rain" is the longest song (at 8:59) to ever make the top ten on the *Billboard* Hot 100 Chart. It was cut down from twenty-five minutes.

Axl Rose wrote "Sweet Child o' Mine" about his girlfriend Erin Everly, who he was married to for one whole month in 1990.

WHO ARE YOU'S?

The Who's *Quadrophenia* album is about a guy named Jimmy who has four personalities (quadrophenia refers to schizophrenia times two), one for each member of the band.

Pete Townshend adopted his trademark windmill arm movements while watching Keith Richards warm up before a show.

SEXZZY SONGS

"Pearl Necklace" by ZZ Top is about oral sex.

"La Grange" by ZZ Top is about a whorehouse in La Grange, Texas—the "Chicken Ranch." It is the same brothel that inspired the movie and play *The Best Little Whorehouse in Texas*. Bassist Dusty Hill lost his virginity there when he was thirteen. The publicity generated by the song forced authorities to close down the cathouse.

YOU GOT ME ON MY KNEES

The Derek and the Dominos hit song "Layla" was inspired by the Persian poem "Layla and Majnun," about two lovers whose parents kept them apart, driving the boy insane.

Duane Allman made that chirping bird sound at the end of "Layla" with his guitar as a tribute to jazz great Charlie Parker, known as "the Bird."

The songs "Layla" by Derek and the Dominos, "Something" by the Beatles, and "Wonderful Tonight" by Eric Clapton are about the same woman—Patti Boyd—who was married to George Harrison and dumped him for Eric Clapton.

PRINCE OF DARKNESS

The Ozzy Osbourne song "Mr. Crowley" was written about early 1900s black magic practitioner Aleister Crowley who was known as "the wickedest man alive." In the song, the normally articulate Ozzy mispronounces the name "Crowley." It should be pronounced with a hard "o," like in "crow."

Crowley is one of the figures pictured on the Beatles' *Sgt. Pepper's Lonely Hearts Club Band* album cover, situated between Mae West and Sri Yukteswar Giri.

HAPPY ENDING

Buddy Holly's "Peggy Sue" was going to be called "Cindy Lou," for Holly's niece. Holly changed the title for drummer Jerry Allison, who had a falling out with his girlfriend, Peggy Sue Gerron. They eventually married.

MATERIAL GIRL

The Madonna song "Vogue" was inspired by "vogueing"—a dance craze in the New York gay community at the time involving posing and elaborate hand gestures to imitate Hollywood stars and *Vogue* magazine covers.

"Borderline" is about an orgasm.

HOLLY, CANDY, AND LITTLE JOE

Lou Reed's "Walk on the Wild Side" is about transvestite prostitutes. The names mentioned in the song—Holly, Candy, Little Joe, Sugar Plum Fairy, and Jackie—were all characters from Andy Warhol's Factory studio.

DON'T WORRY . . .

Bobby McFerrin got the title of his "Don't Worry, Be Happy" from a poster of Indian guru Meher Baba with the phrase on it.

It was the first a cappella song to reach number one on the *Billboard* Hot 100 chart.

SEEING DOUBLE

The Foreigner song "Double Vision" was inspired by guitarist Mick Jones attending a 1977 New York Ranger playoff game, where Ranger goalie John Davidson was knocked out and the PA announced that he was suffering from "double vision." Apparently, this was the first time Jones had heard the expression.

HOLDING PATTERN

In "Arthur's Theme" by Christopher Cross, the line about being "caught between the moon and New York City" was inspired by a time the song's writer, Peter Allen, was stuck in holding pattern above the city, waiting to land at JFK.

Christopher Cross's "Ride Like the Wind" is about a condemned man who flees to Mexico.

I DO RESPECT HER, BUTT

"Her Strut" by Bob Seger was inspired by Jane Fonda's walk in the movie *Barbarella*.

JEREMY SPOKE IN CLASS TODAY

"Jeremy" by Pearl Jam is about a high school student who kills himself to get back at those kids who bullied him. It was inspired by the story of a Richardson, Texas, fifteen-year-old—Jeremy Delle—who killed himself in front of his English class.

ALWAYS ANOTHER SHOW

Journey keyboardist Jonathan Cain wrote "Faithfully," a song about how hard it is to be a married guy in a rock band. He and his wife divorced a few years later.

"I Walk the Line" by Johnny Cash is about his efforts to remain faithful to his first wife, Vivian, while on tour.

Bret Michaels of Poison wrote "Every Rose Has Its Thorn" in a Laundromat after calling his girlfriend from a pay phone on the road and hearing another guy answer.

THAT'S A RAP

"Rapture" by Blondie was the first song with a rap to go to number one on the U.S. *Billboard* Hot 100 Chart. It was also the first rap in a song that had original music, as opposed to rapping over an existing song.

Kid Rock, who was born Robert James Ritchie, began his musical career in rap and hip-hop.

R-E-S-P-E-C-T

Otis Redding wrote and recorded "Respect" before Aretha Franklin made it famous.

LINER NOTES

The "brass" in the Pretenders song "Brass in Pocket" is a Northern England euphemism for money.

The Pretenders' "Middle of the Road" is about middle age.

The song "Unchained Melody" is from the 1955 movie *Unchained*, about a prison inmate who is debating whether to escape and live on the run or serve his time and return to his family.

Metallica's "Until It Sleeps" is about the passing of James Hetfield's mother from cancer.

The Kansas hit "Dust in the Wind" was inspired when guitarist Kerry Livgren read the line in a book of Native American poetry.

Queen's "Radio Ga Ga" was originally titled "Radio Ca Ca" and was a put-down of radio.

"I'm Too Sexy" was a spoof song written by two brothers who made up the group Right Said Fred. It made fun of the vain, macho guys that went to their gym.

"I'm Your Boogie Man" by KC and the Sunshine Band was written in tribute to the Miami DJ who first played their song "Get Down Tonight" on the radio, which was their big break.

Stevie Wonder's "Isn't She Lovely" was written for his daughter Aisha. She can be heard laughing in the song.

Peter Gabriel wrote "Solsbury Hill" about a spiritual experience he had atop the hill in Somerset, England.

"Soul Kitchen" by the Doors is about Olivia's soul food restaurant in Venice Beach, California, that Jim Morrison used to frequent.

"Sara Smile" by Hall & Oates was written for Darryl Hall's girlfriend Sara Allen.

Julie Andrews's "Supercalifragilisticexpialidocious," from the film *Mary Poppins*, was written by Robert and Richard Sherman, who also wrote "Chitty Chitty Bang

Bang" and "It's a Small World." They say they picked up the word at summer camp in the Adirondacks.

"London Calling" by the Clash is about the apocalypse and the many ways that it may occur.

"Love Shack" is about a Georgia cabin that B-52s singer Kate Pierson lived in during the 1970s. The song was taken from the Temptations' "Psychedelic Shack."

"Lucky Man" by Emerson, Lake, and Palmer was written by Greg Lake when he was twelve.

"Who Let the Dogs Out" by the Baha Men is a reference to disrespectful men who hit on pretty girls.

Sting claims it was his son who came up with the title to the Police song "De Do Do Do, De Da Da Da."

Van Morrison's "Domino" is a tribute to Fats Domino.

Jim Morrison recorded the vocals to "L.A. Woman" in a bathroom for a fuller sound.

Ray Manzarek used an electric piano to create the rain sound effect in "Riders on the Storm."

Devo's Mark Mothersbaugh has composed music for Apple commercials and the TV show *Rugrats*.

Matchbox Twenty singer Rob Thomas was homeless for three years after he dropped out of high school and left home at seventeen.

When MTV launched in 1981, only a few thousand people on one cable network in North Jersey could get it.

ARE YOU MY MOTHER?

Eric Clapton was raised by his grandmother and her second husband, who he believed were his mother and father. He thought his real mom, who had him at age sixteen, was really his older sister. Clapton's biological father was a Canadian soldier stationed in England during World War II who had an affair with his mother and went back to his wife in Canada at war's end.

In 2004, Eric Clapton bought a 50 percent share of a failing London clothing store that he had patronized since he was a teen.

"S.O.S." by ABBA is the only top twenty hit song in which the song's title and the group's name are both palindromes—spelled the same way forward and backward.

The Sex Pistols' album *God Save the Queen* went to number two in the UK, despite its being banned there.

"Mickey" by Toni Basil was originally a song named "Kitty," from another group. She liked the song and changed the title to honor the Monkees' Micky Dolenz, who she was quite fond of.

Gerry Rafferty's "Baker Street" was written about Baker Street in London, where a friend of his lived. (It's also the home of the fictional detective Sherlock Holmes.)

Bob Dylan wrote "Blowin' in the Wind" in ten minutes, setting the lyrics to an old slave song.

Peter Wolf, the singer for the J. Geils Band, was married to Faye Dunaway.

Elvis Costello used to be a computer programmer.

Ruby Tuesday restaurants take their name from the Rolling Stones song.

SUPER SALES

In the twentieth century, the Beatles sold the most albums, Elton John's "Candle in the Wind '97" was the bestselling single, and the Eagles' *Greatest Hits 1971–1975* tied with Michael Jackson's *Thriller* as the top-selling albums in the United States.

Styx was the first band to have four consecutive triple-platinum albums.

In 2007, "Rockstar" by Nickelback was the most downloaded music video on iTunes.

NAME THAT BAND

The Dixie Chicks are named after the Little Feat song "Dixie Chicken."

The Doors took their name from Aldous Huxley's *Doors of Perception*, which relates his experiences taking the psychedelic drug mescaline.

Black Sabbath got their name from the 1964 Boris Karloff movie of the same name.

SPREAD THE WEALTH

In 1995, Sting had over one hundred different bank accounts around the globe.

After a 1978 show, Aerosmith bailed thirty of their fans out of jail after they were arrested for smoking marijuana.

MULTITASKER

Iron Maiden singer Bruce Dickinson is a licensed commercial pilot and flew the band's jet on their 2008 world tour.

AU NATUREL

The first time Kiss appeared in public without their makeup was in 1983.

NOT THAT SONG AGAIN!

"My Humps" by the Black Eyed Peas was voted "Most Annoying Song" in a 2010 poll by *Rolling Stone* magazine. Rounding out the top five were "Macarena" by Los Del Rio, "Who Let the Dogs Out" by Baha Men, "My Heart Will Go On" (the *Titanic* song) by Celine Dion, and "Photograph" by Nickelback.

ONE FOGGY CHRISTMAS EVE

Robert L. May, a copywriter for Montgomery Ward, created the story of Rudolph the Red-Nosed Reindeer in 1939 to feature in the coloring book that the company gave away each Christmas. That year, 2.4 million copies were distributed.

May reflected himself in the story. He too was a scrawny misfit as a child.

May's brother-in-law was songwriter Johnny Marks, who set the story to music.

🌰 MISTLETOE MAKE OUT

Jimmy Boyd was twelve years old when he sang the tune "I Saw Mommy Kissing Santa Claus" in 1952. The song has sold an estimated 60 million copies since then.

The record was banned by the Catholic Church in Boston because it mixed Christmas and "sex." After Jimmy went to Boston to explain the song to Church leaders, the ban was lifted.

UNWANTED TRIBUTE

"Only Wanna Be with You" by Hootie and the Blowfish is a tribute to Bob Dylan, who wasn't pleased with the song using lyrics from his work without permission. He sued the group and got a healthy settlement.

THE LOWDOWN

The producers of *Saturday Night Fever* wanted to use Boz Skaggs's "Lowdown" in the film, but his manager opted to have the song used in *Looking for Mr. Goodbar* instead. The *Saturday Night Fever* soundtrack went on to become one of the bestselling albums of all time.

BAD CALL

Meatloaf's "Paradise by the Dashboard Light" doesn't make baseball sense. No team would try a suicide squeeze with two outs.

BORN ON CHRISTMAS DAY

Like the character in his 1971 hit song "Levon," Elton John's son was born on Christmas Day, in 2010. The boy's name is Zachary Jackson Levon Furnish-John. Elton had the baby with longtime partner David Furnish by means of a surrogate mother. Bernie Taupin named the song after Levon Helm, the drummer and singer for the Band, he and Elton's favorite group at the time.

THANKS, MOM

Whitney Houston's mother, Cissy Houston, sings backing vocals on "How Will I Know?"

ROWDY ROCKERS

Johnny Ramone, guitarist for the Ramones, had emergency brain surgery to remove blood clots after being repeatedly kicked in the head in a 1983 fight.

In 1998, Aerosmith drummer Joey Kramer received second-degree burns when his car caught fire while he was pumping gas.

Deep Purple wrote "Smoke on the Water" about a fire that destroyed a casino in Montreux, Switzerland, just before they were scheduled to record their *Machine Head* album. The blaze was started during a Frank Zappa and the Mothers of Invention show at the venue.

The heroin that Sex Pistols' bass player Sid Vicious overdosed and died on was supplied by his mother.

Marc Bolan, singer and guitarist for T. Rex, died in a car wreck in 1977 when his girlfriend plowed into a tree in London.

David Lee Roth slashed his forehead accidentally while doing a samurai sword schtick onstage in 2003. He needed twenty-two stitches and cancelled the rest of the dates on his tour.

Jimi Hendrix died from choking on his own vomit after swallowing nine of his girlfriend's sleeping pills and drinking wine.

Beach Boys guitarist Carl Wilson was indicted in 1967 for refusing to be sworn into the army after being drafted.

Beach Boys drummer Dennis Wilson put his hand through a windowpane in 1971, cutting himself so badly that he couldn't play for three years.

In 2004, Eric Clapton was banned from driving in France after he was clocked doing 134 miles per hour in his Porsche 911 Turbo.

Sting fired his personal chef because she became pregnant (not by him). He later was ordered by the court to pay her fifty thousand dollars for sexual discrimination.

Mötley Crüe singer Vince Neil was convicted of vehicular manslaughter for driving his Pantera sports car head-on into an oncoming vehicle while drunk in 1984. The wreck killed his passenger, Hanoi Rocks drummer Nicholas "Razzle" Dingley, and severely injured the two occupants of the other car. He spent twenty days in jail and paid $2.6 million in compensation.

At a 1971 concert in London, Frank Zappa was pushed off the stage by a jealous boyfriend. He suffered a broken leg and cracked skull, which kept him wheelchair-bound for nine months.

In 1990, Billy Idol put six hundred dead fish in the dressing room of the band Faith No More. As payback, they streaked across the stage during Idol's set.

Ike Turner reportedly spent $11 million on his cocaine habit, before cleaning up his act.

Atlantic Records founder Ahmet Ertegun died a few weeks after falling at a Rolling Stones concert and hitting his head in 2006. He was eighty-three.

In 2002, Alice in Chains singer Layne Staley died of a drug overdose in his apartment. His body was not discovered for two weeks.

Judas Priest was banned from Madison Square Garden after fans ripped out seat covers and threw them onstage during a 1984 concert, causing $250,000 in damage.

In 1996, Jimmy Buffett's plane, which was also carrying Bono, was fired upon by authorities in Jamaica who thought it belonged to drug smugglers.

Ray Davies of the Kinks was shot in the leg in 2004 after chasing a purse snatcher who had grabbed his girlfriend's bag.

Jimi Hendrix once set fire to his guitar onstage, burning his hands and requiring hospitalization.

A trip to the hospital after burning himself with hot water inspired Joey Ramone to write "I Wanna Be Sedated."

NO BRAINER

Kurt Cobain and Courtney Love named their child Frances Bean after Frances Farmer, a lobotomized 1940s actress.

IN CONCERT

In 1995, Def Leppard did three shows on three continents in one day—one in London, one in Tangiers, Morocco, and one in Vancouver, Canada.

The Rolling Stones' 1981 Tattoo You Tour was the first tour to have a corporate sponsor.

The 1968 Newport Pop Festival in Costa Mesa, California, was the first rock concert to draw one hundred thousand people.

During a 2010 concert at the Verizon Amphitheatre in St. Louis, the Kings of Leon were forced to leave the stage after three songs due to unruly pigeons that kept pooping on the band from above.

ON SECOND THOUGHT

Joan Jett's "I Hate Myself for Loving You" was originally titled "I Hate Myself Because I Can't Get Laid." She had to change it for obvious reasons.

Chuck Berry changed the line "That little colored boy can play" to "That little country boy can play" so "Johnny B. Goode" would be acceptable for radio play.

Van Morrison's hit "Brown Eyed Girl" was originally titled "Brown-Skinned Girl."

IDOL THOUGHTS

PURELY PADMA

Top Chef host Padma Lakshmi was born in India and raised a vegetarian.

Lakshmi has a large scar on her arm from a car wreck that she was in when she was fourteen. She says that the accident is "one of the most beautiful images in my memory."

Lakshmi was married to author Salman Rushdie from 2004 to 2007.

Lakshmi speaks English, Hindi, Italian, Spanish, and Tamil.

RECIPE FOR DISASTER

Two of the chefs that have appeared on *Kitchen Nightmares* with Gordon Ramsay have since committed suicide.

BEAVER TALES

Barbara Billingsley, the mother on *Leave It to Beaver*, always wore pearls to hide an unsightly indentation just above her sternum.

The incorrigible character Eddie Haskell was voted number two on *TV Guide*'s list of all-time TV brats.

LIVE FROM NEW YORK . . .

Saturday Night Live was broadcast with a five-second delay three times—when Andrew Dice Clay, Sam Kinison, and Richard Pryor each hosted.

SNL was initially intended to run just six episodes.

Mike Myers based his *SNL* character Dieter on an acquaintance from art school. He based the Linda Richman character on his mother-in-law.

Carly Simon taped her musical performance in 1976 because she suffered from stage fright. ABBA lip-synched their 1975 appearance.

Ashlee Simpson was greatly embarrassed in 2004, when the wrong song was played and the lyrics began before she even raised the microphone to her mouth to begin lip-synching.

Paul Schaffer was the first person to accidentally use the F-word live on *SNL* in 1980.

The *SNL* dress rehearsals are all taped and occasionally certain broadcast sketches are replaced with the rehearsal bits for re-airing, if obscenities or technical difficulties occur.

While Larry David was a writer for the show, he barged into executive producer Dick Ebersol's office and quit. He soon realized what a big mistake that was and just showed up for work again the next day, like nothing had happened. David used this experience on the *Seinfeld* episode "The Revenge," where George does the same thing.

Conan O'Brien appeared many times on the show uncredited during his stint as an *SNL* writer from 1988 to 1991.

John Goodman unsuccessfully auditioned for the show in 1980. He came back to guest host eleven years in a row.

"SHOOTING" STAR

John Belushi was born in Chicago of Albanian parents.

In 1979, on Belushi's birthday, he had the number one movie (*Animal House*), the number one album (*The Blues Brothers: Briefcase Full of Blues*), and was on the number one late night TV show (*Saturday Night Live*).

Belushi could drink a fifth of Jack Daniel's in five minutes.

He died in 1982 after being injected with eleven speedballs (a combination of cocaine and heroin) administered by Cathy Smith, a band groupie and drug dealer. She was sentenced to fifteen months in prison for involuntary manslaughter.

Robin Williams and Robert DeNiro both had visited with Belushi the night of his death.

TOGA! TOGA!

The movie *Animal House* was filmed in and around the University of Oregon in Eugene.

Toga parties became the rage after the film's release.

John Belushi's exploding zit scene in the cafeteria was improvised.

Universal Pictures president Ned Tanen wanted the scene at the African American bar removed because he thought it would cause race riots in theaters. After Richard Pryor saw a screening of the film and loved the scene, it was decided to leave it in.

Donald Sutherland thought the movie was so bad that he took a flat $75,000 fee for his three days' work, instead of a gross percentage that would have made him around $4 million.

Animal House was supposed to be set in Pennsylvania. When filming the courtroom scene, the set decorator could not find a Pennsylvania flag, so used a Tennessee flag instead.

Meatloaf was the director's second choice to play John Belushi's character Bluto, in case Belushi dropped out.

ENDORSED BY . . .

Many stars aren't content with just being stars. Some also market their own unique products:

> Aerosmith guitarist Joe Perry has his own hot sauce—Boneyard Brew.

Hulk Hogan has Hulkster frozen cheeseburgers. Hogan has stated that his agent was approached for him to endorse two products—a grill and a blender. After Hogan missed a phone call, George Foreman signed on to lend his name to the grill. Hogan got stuck with the blender, which went nowhere.

> Rolling Stone Bill Wyman, who is into archaeology, sells his own signature metal detector.

Sylvester Stallone has Stallone High Protein Pudding.

> Steven Seagal has Steven Seagal's Lightning Bolt energy drink.

Carlos Santana has a line of high-end shoes.

> Sophia Coppola has her name on canned sparkling wine made by her father's (Francis Ford Coppola) vineyard.

Many of the scores of products that Jackie Chan endorses in China seem to have been cursed, including an anti-hair-loss shampoo that contains carcinogens; the auto repair school that was involved in a diploma scandal; the compact disc player maker that went bankrupt; and the air conditioners that exploded.

That perky Progressive car insurance girl with the hair bump on her head is named Flo. She is played by comedienne Stephanie Courtney.

SWEATY STARS

Some rather curious stars have had their own workout videos, including Joan Lunden, Mary Tyler Moore, Regis Philbin, Zsa Zsa Gabor, Traci Lords, and Barbie.

SALES STARS

Many big stars got their starts doing commercials:

Elijah Wood hawked Pizza Hut.

Leonardo DiCaprio blew bubbles for Bubble Yum.

Meg Ryan flashed her pearly whites for Aim toothpaste.

Whitney Houston sang for Canada Dry ginger ale.

John Travolta sang about Lifebuoy soap while taking a shower.

Morgan Freeman extolled the virtues of Listerine from the top of a telephone pole.

Matt LeBlanc enjoyed Heinz ketchup.

Dustin Hoffman pushed Volkswagen.

Susan Sarandon let three globs of lotion run down her hand for Dermassage.

Seth Green was in a Nerf commercial.

Keanu Reeves appeared in a Corn Flakes ad.

Evangeline Lilly of *Lost* once did TV commercials for Canadian local singles phone chat lines.

DIDN'T YOU USED TO BE . . .

Jamie Foxx's given name is Eric Marlon Bishop.

Liv Tyler was born Liv Rundgren. At age thirteen, she changed her name after figuring out that Aerosmith singer Steven Tyler was her father, not singer Todd Rundgren.

Jenna Jameson used to be Jenna Marie Massoli. She picked the name Jameson out of the phone book.

Ben Kingsley was born Krishna Bhanji. He wanted his name to sound more British to get acting roles.

Clay Aiken began life as Clayton Grissom. Aiken is his mother's maiden name.

Demi Moore used to be Demi Gene Guynes. She kept the name Moore from her marriage to Freddy Moore in 1980.

Elle Macpherson was once Eleanor Nancy Gow.

Ozzy Osbourne was christened John Michael Osbourne.

Bret Michaels ditched his given surname—Bret Michael Sychak.

David Copperfield thought better of his given name—David Sethkotkin.

Diane Keaton was born Diane Hall.

Wynonna Judd entered this world as Christina Claire Ciminella.

John Cleese was born John Cheese.

Tori Amos began life as Myra Ellen Amos.

Joni Mitchell used to be named Roberta Joan Anderson.

Sheena Easton was Sheena Shirley Orr until her marriage to Sandi Easton in 1979.

Fergie is a play on her given name—Stacy Ann Ferguson.

Alicia Keys began life as Alicia Augello Cook.

Barry Manilow thought better of his birth name—Barry Alan Pincus.

Flavor Flav was born with the rather bland name William Jonathan Drayton Jr.

Pink didn't think her original name—Alecia Beth Moore—was colorful enough.

Meg Ryan wasn't sweet on Margaret Mary Emily Anne Hyra.

Ice-T didn't think Tracy Marrow was cool enough.

Courtney Love didn't love Courtney Michelle Harrison.

Carmen Electra wasn't charged up about Tara Leigh Patrick.

Steven Tyler didn't think Stephen Victor Tallarico rocked.

FAMILY AFFAIRS

Kiefer Sutherland and his father, Donald, both appeared in the 1983 film *Max Dugan Returns* and 1996's *A Time to Kill*.

Michael Douglas and his father, Kirk, both appeared together in the 1966 movie *Cast a Giant Shadow*.

Brother and sister Julia and Eric Roberts were both in the 1987 film *Blood Red*.

Angelina Jolie and her father, Jon Voight, worked together in the 1982 movie *Lookin' to Get Out* and again in the 2001 flick *Lara Croft: Tomb Raider*.

FOR STARTERS

Cameron Diaz made her movie debut in the 1994 film *The Mask*.

Drew Barrymore started appearing in TV commercials before she was one year old. At age five she appeared in the 1980 movie *Altered States*. One year later, she starred in the blockbuster *E.T.*

Before George Clooney made it big in movies, he "starred" in 1987's *Predator: The Concert* and *Return to Horror High*, as well as *Return of the Killer Tomatoes* in 1988.

Jodie Foster made her TV acting debut on an episode of *Mayberry R.F.D.* in 1968.

Foster also appeared on *Gunsmoke*, *Adam-12*, and was one of Eddie's friends on *The Courtship of Eddie's Father*.

Sarah Jessica Parker played Annie on Broadway when she was fourteen.

Amy Adams used to be a Hooters girl.

JUST DUCKY

The Aflac duck "speaks" more softly in Japanese commercials and the other people in the commercial don't ignore him, which would be considered rude in Japan.

Comedian Ray Romano was originally considered to promote Aflac, but lost out to the duck in testing.

TOTAL RECALL

Taxi star Marilu Henner is one of only seven people in the world identified with hyperthymesia, or Superior Autobiographical Memory. She can recall events from every day of her life.

TURNER TIME

Lana Turner's fourteen-year-old daughter, Cheryl Crane, stabbed Turner's mobster boyfriend Johnny Stompanato to death in 1958. It was ruled a case of justifiable homicide since Stompanato was very abusive to Turner.

Stompanato once pointed a gun at Sean Connery on the set of a movie he was doing with Turner, out of jealousy. Connery disarmed him and beat him up before ejecting Stompanato.

MILLION-DOLLAR MAN

Comedian and actor Roscoe "Fatty" Arbuckle was the first star to earn a million dollars a year.

In 1920, Arbuckle was accused of raping a woman in a hotel room so violently that she died of internal injuries. After three manslaughter trials, Arbuckle was finally found not guilty, but he had already been banned from Hollywood and his career was ruined.

OLD FLAMES

Matt Damon used to date Winona Ryder.

Brad Pitt went out with Juliette Lewis.

Madonna and Sean Penn were married for four years.

Meg Ryan and Russell Crowe once had a thing.

Demi Moore was engaged to Emilio Estevez.

Sarah Jessica Parker went with Robert Downey Jr. for seven years, before marrying Matthew Broderick.

AND THE WINNER IS . . .

The Antoinette Perry Awards for Excellence in the Theatre are more commonly known as the Tony Awards. Antoinette Perry was an actress, director, producer, and cofounder of the American Theatre Wing, the group that started the awards in 1947. Her nickname was "Toni."

The musical that won the most Tony Awards was 2001's *The Producers*, with twelve.

Director/producer Harold Prince has won the most Tonys, with twenty-one.

Julie Harris and Angela Lansbury have won the most performing awards, with five each.

DOES THAT NAME COME WITH COUNSELING SESSIONS?

Many stars feel compelled to give their offspring creative names. Some notable examples follow:

Toni Braxton's kids are Denham and Diesel.

Nicolas Cage's child is Kal-El.

Anne Heche's two progeny are Atlas and Homer.

Richard Gere also has a Homer—Homer James Jigme.

Rob Morrow's daughter is Tu Morrow (as in "the sun will come out tomorrow").

Forest Whitaker's quartet are Autumn, Ocean, Sonnet, and True.

Bob Geldof's girls are Fifi Trixibelle, Little Pixie, and Peaches Honeyblossom.

The Edge fathered Blue Angel.

Penn Jillette sired Zolten and Moxie Crimefighter.

Gwen Stefani gave birth to Zuma Nesta Rock.

Ashlee Simpson is the mother of Bronx Mowgli.

Angelina Jolie has three biological kids—Shiloh, Knox, and Vivienne—and three adopted kids—Maddox, Pax, and Zahara.

CARRIE ON

Sex and the City was loosely based on the *New York Observer* columns of writer Candace Bushnell and her book of the same name. She created the character Carrie Bradshaw in her essays, who shares her own initials.

Sex and the City was the first cable television series to win a Grammy for Best Comedy.

Cynthia Nixon, who played Miranda, is a natural blonde and had to dye her hair red for the show. Her earrings are all clip-ons, as her ears are not pierced.

Sarah Jessica Parker had a clause in her contract that she would not do any full nudity, unlike her costars.

The first name of Mr. Big (John) is not revealed until the final episode of the series. His last name (Preston) wasn't revealed until the first movie came out.

The last word of the last episode, spoken by Carrie, is "fabulous."

CELEBRITY CONFIDENTIAL

Vanessa Williams used to apply her own fresh urine to her face to treat acne.

Heidi Montag once had ten plastic surgery procedures in one day.

Jessica Simpson has been known to openly fart in business meetings.

Sienna Miller likes cigarettes so much that she has been quoted as saying, "I think the more positive approach you have to smoking, the less harmful it is."

Nicolas Cage owns a nine-foot-tall pyramid tomb in New Orleans that he plans to spend eternity in.

Angelina Jolie aspired to be a funeral director in her teens. She also collected knives and liked to cut herself.

ACT YOUR AGE

Many older stars have played teenagers over the years:

Sissy Spacek was in her late twenties when she played the troubled supernatural teen in the film *Carrie*.

Michael J. Fox was twenty-four when he played teenage Marty McFly in *Back to the Future*.

Jon Heder was in his late twenties when he played Napoleon Dynamite.

Olivia Newton-John was twenty-nine when she played the sweet teen Sandy in *Grease*.

Several members of the cast of *Beverly Hills 90210* were in their mid- to late twenties in this teen television show.

Gary Burghoff was twenty-nine when he first played the eighteen-year-old Corporal Radar O'Reilly in the TV show *M*A*S*H*.

CANDLE IN THE WIND

Princess Diana dropped out of high school at age sixteen after failing all of her O-level exams (tests given to sixteen-year-old students in Britain) twice. She then went to finishing school in Switzerland, where she also dropped out.

Diana worked as a preschool assistant and did cleaning work for her sister and friends.

Di first met her future husband, Prince Charles, when he was dating her older sister, Lady Sarah.

Charles gave Diana a £30,000 engagement ring, which their son Prince William gave to his fiancée, Kate Middleton.

The train on Diana's wedding dress was twenty-five feet long.

Diana messed up her wedding vows, omitting the word "obey" and stating Charles's full name in the wrong order.

Charles and Di's fairy-tale wedding was watched on television by some 750 million people worldwide.

BONDAGE

Ian Fleming originally wanted David Niven or Roger Moore to play 007 in the first Bond films. The producers preferred Sean Connery. However, Moore got his chance after Connery later stopped playing Bond.

FIVE-STAR ENTERTAINER

Gene Autry is the only person with five stars on the Hollywood Walk of Fame—one each for movies, radio, music recording, TV, and live theater.

YOU CAN'T TAKE IT WITH YOU

In 2010, Michael Jackson made the most money of any dead celebrity—$275 million.

RICH WITCH

Emma Watson had only acted in school plays before she was cast as Hermione in the first Harry Potter movie at age nine.

Watson only received an allowance of $75 a week when she was seventeen, despite being worth more than £10 million.

In 2010, Watson was Hollywood's highest-paid actress, worth about $32 million.

Rupert Grint, who plays Ron Weasley in the Harry Potter movies, won the role by sending in an unsolicited video to the producers. He too had only acted in school plays before this.

OUTTAKES

Will Rogers was 9/32 Cherokee.

Eric Stolz played the part of Marty McFly for the first five weeks of the shooting of *Back to the Future*. Stolz was replaced with Michael J. Fox, because he wasn't funny enough.

In the movie *Harry Potter and the Sorcerer's Stone*, Fiona Shaw, who played Aunt Petunia, had dead mice hung from her apron in one scene so that the owls would look at her.

Homer Simpson attends the First Church of Springfield and is a "Presbylutheran."

Jennifer Aniston is the godmother of Courteney Cox's daughter, Coco.

Courteney Cox has a brown belt in karate.

Miramax Films was founded by Bob and Harvey Weinstein, who named the company for their parents, Miriam and Max.

Meredith Baxter's mother, Whitney Blake, played the mother on the TV show *Hazel*.

The exterior shots of Jerry's apartment on *Seinfeld*, which look nothing like Manhattan, are of a building at 575 New Hampshire Avenue in Los Angeles.

MOVIE MOMENTS

The first full-length movie shown in America was 1913's *Queen Elizabeth*, starring Sarah Bernhardt.

The first full-length talking motion picture was 1927's *Jazz Singer*, starring Al Jolson.

TATTOO YOU

Many celebs have tattoos that they came to regret, or may do so one day:

Heather Locklear had the "Richie" tattoo on her hip, which was to honor her husband Richie Sambora, changed to a rose after they split up.

Pamela Anderson had the "Tommy" tattoo on her ring finger, which she got for Tommy Lee, altered to read "Mommy." She kept the barbed wire tat around her bicep.

Denise Richards had the "Charlie" tattoo, which she had done for Charlie Sheen, turned into a fairy.

Britney Spears still has a pink pair of dice inked on her left wrist that match a blue pair of dice tattooed on ex-husband Kevin Federline's right wrist.

Angelina Jolie underwent painful laser treatments to remove her "Billy Bob" tat from her upper arm that she got for Billy Bob Thornton.

Tom Arnold had four tattoos featuring Roseanne Barr removed after they broke up.

Dean McDermott has wife Tori Spelling's face and bikini-clad boobs inked on his upper arm, as well as three other Tori Spelling tattoos.

Steve-O, of *Jackass* fame, has a full back tattoo honoring himself.

Lest she forgets to do so, Lindsay Lohan has "breathe" inked on her wrist.

Hayden Panettiere has a misspelled tattoo in Italian that is supposed to read *"Vivere senza rimpianti,"* which means "To live with no regrets." What it actually reads is *"Vivere senza rimipianti,"* which means absolutely nothing.

Pink has a bar code tattooed on the back of her neck.

Amanda Seyfried is proud of the tattoo of her nickname "Minge." It is a British euphemism for "vagina." The artwork, however, is located a little lower down, on her foot.

Rihanna has a tattoo that reads *"rebelle fleur,"* which is supposed to be French for "rebellious flower." However, to be grammatically correct it should read *"fleur rebelle."* She also has one on her index finger that reads "Shhh . . ."

DIVA DIRT

Vogue magazine editor in chief Anna Wintour, inspiration for the domineering editor in the book and movie *The Devil Wears Prada,* is infamous for not riding on elevators with other people. She once had her bodyguards carry her down five flights of stairs at a New York fashion show to avoid such an ordeal.

Martha Stewart has a real problem with Diet Coke. She prohibits it to be anywhere near her and lectures her television audiences about its evils.

Fashion designer Isaac Mizrahi once went into a tirade because a security guard at an office he was visiting had on a brown uniform.

WRITE ON

LOVE'S LABOUR WON

William Shakespeare may have written a sequel to *Love's Labour's Lost—Love's Labour's Won*. This title was listed as one of his works at the time, but no one knows if it was a lost play or simply an early title of another play.

WHAT THE DICKENS?

When Charles Dickens was just twelve, his father was sent to debtors prison and Charles had to drop out of school and work ten-hour days in a shoe polish factory, where he pasted labels on jars.

RULE BRITANNICA

Encyclopaedia Britannica was managed for eighteen years (1920–23 and 1928–43) by Sears Roebuck. Until Sears took over, the encyclopedia was only updated once every twenty-five years.

WRITER'S CRAMP?

The bestselling children's author of all time is R. L. Stine, who writes about two novels every month and has sold more than 400 million books.

Ernest Hemingway wrote 500 words a day.

WRITER'S TRIPLE CROWN

In 1994, Michael Crichton became the only writer in history to have had the number one movie (*Jurassic Park*), the number one TV show (*ER*), and the number one book (*Disclosure*) simultaneously.

Crichton was trained as both a physician and a physical anthropologist at Harvard University.

BUY ME A MOCKINGBIRD

Harper Lee's first and only novel, *To Kill a Mockingbird*, published in 1960, has sold more than 30 million copies. It is the most widely read novel in high schools today and still sells about 1 million copies a year.

The *Library Journal* voted *Mockingbird* "Best Novel of the Century" in 1999.

The character Scout is modeled on Lee and the character Dill is based on Lee's childhood friend, Truman Capote.

Lee helped Capote do research for his 1966 book *In Cold Blood*.

Harper Lee basically became a recluse after achieving her fame.

AMERICAN LIT

Eugene O'Neill is the only American playwright to have won a Nobel Prize for Literature.

Walt Whitman only received formal schooling until the age of eleven, after which time he began working as a newspaper apprentice.

Ralph Waldo Emerson enrolled in Harvard University when he was fourteen.

John Steinbeck dropped out of college and worked at a fish hatchery before he became a successful writer.

William Faulkner stood five feet five inches tall.

Madonna's children's book—*The English Rose*—debuted at number one on the *New York Times* best-sellers list in September 2003.

POE-TENT PROSE

Edgar Allan Poe had a keen interest in cryptography and helped to popularize cryptograms in newspapers.

Poe's first published book only sold fifty copies.

Poe's first success came with the publication of his narrative poem—*The Raven*. Although he only received nine dollars for the work, it made him an overnight sensation.

GO ASK ALICE—TO TAKE HER CLOTHES OFF

Lewis Carroll enjoyed taking nude photographs of young girls.

A POCKETFUL OF BONERS

The following book titles can be very misleading:

Drummer Dick's Discharge (A 1902 children's novel about a drummer boy who gets discharged from the military.)

The Pocket Book of Boners (A 1943 book of jokes illustrated by Dr. Seuss.)

Games You Can Play with Your Pussy (A 1985 book, subtitled . . . *and Lots of Other Stuff Cat Owners Should Know*.)

The Day Amanda Came (A children's book published in 1971.)

FOR THE BIRDS

John James Audubon illustrated the greatest bird guidebook in history—*The Birds of America*—and today it is the most valuable book at auction, but when he produced his masterwork, no publisher in the United States would touch it. He had to take it to England, where it was enthusiastically received.

MOLDY MANUSCRIPT

The first known published book was the Diamond Sutra, in China, in AD 868.

CALL ME UNSUCCESSFUL

Moby Dick, which many consider to be the "Great American Novel," received little recognition at the time of its publication in 1851. It wasn't until thirty years after author Herman Melville's death that it gained notoriety.

Melville wrote a sixteen-thousand-line poem—*Clarel*—that bombed. All unsold copies were burned when he could not afford to buy them at cost.

Melville earned just ten thousand dollars from his writing during his lifetime.

Fifteen years before Melville died, all his books were out of print.

MARK HIS WORDS

Before adopting the pen name Mark Twain, Samuel Clemens wrote under the name Thomas Jefferson Snodgrass.

Mark Twain formed a Confederate militia during the Civil War known as the Marion Rangers. They disbanded after two weeks.

Mark Twain started a publishing company that published the *Personal Memoirs of Ulysses S. Grant*, a huge bestseller, which made Twain rich. However, other memoirs he published (notably one on Pope Leo XIII that sold just two hundred copies) bombed and he lost a lot of money.

> Twain wrote most of his best-known books in Hartford, Connecticut.

BOOK BRIEFS

There really was a Hunchback of Notre Dame. He worked at the Cathedral of Notre Dame around the same time that Victor Hugo wrote his famous story (1828–31).

> Theodore Seuss Geisel, better known as Dr. Seuss, did an ad campaign for an insect spray called Flit. His bug cartoons were featured in the ads for seventeen years.

Mallanaga Vātsyāyana, the Hindu philosopher who supposedly authored the *Kama Sutra*, was celibate.

> The first paperback book to sell 1 million copies was Dale Carnegie's *How to Win Friends and Influence People* in 1936.

BOOK NOOKS

William Barnes and G. Clifford Noble started Barnes & Noble when they opened their first bookstore at 31 West 15th Street in New York in 1917.

Today there are more than seven hundred Barnes & Noble stores in the United States.

Random House is the largest English-language publisher of trade books. The Penguin Group is number two.

DARE TO EXPLORE

The National Geographic Society published its first issue of *National Geographic* magazine in 1888. It was sent to the society's 165 members and contained mostly reprints of papers presented at society meetings.

The first photograph of a natural scene published in *National Geographic* was one of Herald Island, Alaska, in 1890.

In 1897, Alexander Graham Bell was elected president of the National Geographic Society.

BOOKWORTHY

Abraham Lincoln is the American president who has had the most books written about him—2,140. George Washington has been the subject of 1,300; Thomas Jefferson, 882; George W. Bush, 692; and John F. Kennedy, 677.

WORDWISE

In England, "meow," the sound a cat makes, is spelled "miaow."

Ulysses is the Latin name for Odysseus, the character from Greek mythology.

JCPenney used to be known as J.C. Penney. In 2011, they became jcpenney.

Bullfighting is also known as tauromachy.

The word "bimbo" derives from the Italian word for male baby, *bambino*. This makes sense because the earliest English usage (circa 1919) of the word was to describe a clueless man. Bimbo did not come to be associated with women until 1929.

The word "glitch" didn't enter the lexicon until the early 1960s. It was first used by space engineers when referring to a temporary system fault. The word comes from the German *glitschig*, meaning "slippery."

Confetti is a sugarcoated almond candy given out at weddings in Italy.

The word "paparazzi" first appeared in the 1960 Federico Fellini movie *La Dolce Vita*. One of the characters in the film is news photographer Paparazzo.

"Go-Go" derives from the French *à gogo*, meaning in "abundance, galore."

The fedora was originally a woman's hat. It was introduced by actress Sarah Bernhardt in the 1882 play—*Fédora*—where she played a princess who wore a similar style hat. It wasn't until the early twentieth century that the fedora became a man's hat.

"Beats me," "blue chip," "when the chips are down," and "pass the buck" are all expressions derived from poker.

The word "pope" comes from the Greek *pappas*, meaning "father."

The small guitar-like instrument that became the ukulele was introduced to Hawaii from Portugal in 1879. Called a *machete* by the Portuguese immigrants, the little four-stringed instrument quickly won over the locals, who began to call it the ukulele, meaning "jumping flea" in Hawaiian, because of the way a player's fingers jump up and down the fret board.

The word "hotmail" is derived from "html," which stands for "hypertext markup language."

The first limousine was built in the Limousin region of France, hence the name.

The word "galoshes" comes from bad-weather boots worn by the ancient Gauls, which the Romans called "Gaulish boots."

The word "spandex" is an anagram of the word "expands."

The word "angora" comes from the old word for Ankara, Turkey.

The word "satin" comes from the old Chinese seaport—Zaytun.

EPCOT is short for Environmental Prototype Community of Tomorrow.

The word "hacker" originally meant "someone who makes furniture with an ax."

Those colorful balls of fluff that cheerleaders shake are properly known as "pompons," not "pom-poms."

AD (*anno Domini*, "the year of our Lord," in Latin) goes before the year (AD 50), but after the century (fifth century AD).

The plural of attorney general is attorneys general.

The word "googol," the large number that is a 1 with one hundred zeros (and the source of the name "Google"), was coined in 1938, by nine-year-old Milton Sirotta, the nephew of American mathematician Edward Kasner.

The word "penis" comes from the Latin word for "tail."

The word "modem" is short for "modulate/demodulate."

The word "pixel" is short for "picture cell."

"Keelhauling" was a form of punishment used by the Dutch Navy from the mid-1500s until 1853. It involved tying a sailor to a rope that was looped beneath the ship, throwing him over one side, and pulling him under the keel so that the barnacles on the ship's bottom badly cut him. Drowning often resulted.

The official name of Amtrak is the National Railroad Passenger Corporation.

The name "Unabomber" is short for "University and Airline Bomber."

The petri dish is named for German bacteriologist Julius Richard Petri.

Soda "jerks" got their name from the fact that they had to jerk the soda fountain handle back and forth a few times to mix the flavored syrup in the glass into the carbonated water.

The word "hurricane" comes from the Carib word *Huracan*, meaning "god of evil."

As late as the late 1800s, congressmen in Washington lived in boardinghouses or hotels and would spend a lot of time in the lobbies socializing. Ulysses Grant coined the term "lobbying" to describe the mingling of elected officials with industry interest groups in this setting.

The expression "barking up the wrong tree" comes from coonhounds, which are bred to chase their quarry up a tree and bark until the hunter arrives.

GT stands for *gran turismo*, which is Italian for "grand tourer," a fast car made for comfortable, long-distance driving.

The Hottentots of Southern Africa got their name from early Dutch explorers because of the unique clicking sounds they make. *Hottentot* is Dutch for "stuttering."

Sheetrock is a trademark of USG Corporation.

Pilates also is a trademarked name.

Reebok International Limited, makers of athletic shoes, got their name from the African antelope of the same name.

HOTSPOTS

Wi-Fi is a trademark of the Wi-Fi Alliance, a global group of companies that promotes wireless usage.

The term "Wi-Fi" was created by a consulting firm in 1999 at the behest of the alliance. Many believe it stands for "wireless fidelity," although the alliance, which uses these words in press releases, contends Wi-Fi means nothing at all.

POINT OF FACT

The Dewey Decimal System was created by Melvil Dewey in 1876.

Dewey created his book classification system while a librarian at Amherst University by using decimal numbers with a structure of classification devised by Sir Francis Bacon.

The Dewey Decimal System organizes books into ten main classes, with ten divisions and ten sections—that is, ten classes, one hundred divisions, and one thousand sections. Each book has a unique number.

American fiction is numbered 813.

The Dewey Decimal System is proprietary and is used in two hundred thousand libraries in 135 countries.

LANGUAGE ARTS

The first alphabet was developed by an unknown Semitic people between 1800 BC and 1000 BC. It contained twenty-two consonants and no vowels. The reader had to infer the appropriate vowel sound from their knowledge of the language.

A "new" language was discovered in 2010 in northeastern India. Known as Koro, it is spoken by about one thousand people.

English is the dominant language on the Internet. Seventy percent of home pages are in English.

Students in China must start learning English in the third grade. By comparison, only 4 percent of American secondary schools even offer Chinese.

PICTURE THIS

SPLASH AND DASH

Jackson Pollock was expelled from two different high schools.

Pollock used house paints when creating his famous splash and dribble painting instead of artist paints because they were much cheaper.

Pollock died while driving drunk in 1956 at age forty-four.

DANDY ANDY

Andy Warhola is better known as Andy Warhol. His nickname was "Drella," a combination of Dracula and Cinderella.

Warhol had chorea (an abnormal movement disorder) as a child and became a hypochondriac with a fear of hospitals.

Warhol started out as a commercial illustrator.

He produced the Velvet Underground's first album.

Warhol, who didn't drink, claimed he used Absolut vodka as a perfume.

Warhol was shot by a radical feminist in 1968 and almost died.

ARTIST'S GALLERY

El Greco's real name was Doménikos Theotokópoulos.

Paul Gauguin worked as a stockbroker.

Grant Wood used his dentist and sister as models for his painting *American Gothic*.

The Carpenter Gothic–style house depicted in *American Gothic* still stands today in Eldon, Iowa.

GOT IT COVERED

The *Saturday Evening Post* forbade Norman Rockwell from depicting blacks in his cover illustrations, unless they were shown in subservient roles.

When Rockwell later started doing cover illustrations for *Look* magazine, he began a classic series of pro-civil-rights-themed paintings featuring African Americans.

Unlike many artists, Rockwell became a world-famous celebrity during his lifetime and cashed in, doing television commercials for several products, including Zenith TVs and Purina Cat Chow.

FIRST IMPRESSIONS

The Impressionists painters got their name from the Claude Monet painting—*Impression, Sunrise*. French art critic Louis Leroy inadvertently coined the word in a satirical review of the work in 1874.

Vincent van Gogh was a post-Impressionist, not an Impressionist. Post-Impressionists were more likely to distort form and use unnatural colors.

STROKES OF GENIUS

In his early twenties, Vincent van Gogh worked as a preacher in Belgium.

Vincent was a self-taught artist.

There are 1,753 paint strokes in Van Gogh's 1888 masterpiece *The Bedroom*.

Forgers are not able to re-create Van Gogh's paintings in so few brushstrokes, making fakes easy to identify.

There is some evidence that Van Gogh's friend and fellow artist Paul Gauguin may have been the one to cut off Vincent's ear with a sword during an argument.

Vincent presented his severed ear to a prostitute named Rachel.

Van Gogh shot himself in the chest in a field, before walking back to an inn, where he died two days later.

PROLIFIC PABLO

Picasso produced some fifty thousand works of art in his lifetime. He is ranked as the top-selling artist at auctions.

More of Picasso's paintings have been stolen than any other artist.

TAINTED ART

The Louvre opened in 1793. Most of the original pieces exhibited were works seized from the Church and royalty by Napoleon's army. Further artwork was appropriated by Napoleon from the countries he conquered. After his defeat at Waterloo, many of these pieces were returned to their original owners, despite the Louvre's efforts to hide them.

Hundreds of works of art stolen by the Nazis from Jews and others during World War II ended up in the Louvre, where many remain today, in spite of calls for their return to their rightful owners.

BEWARE OF CATS

The Hermitage art museum was founded by Catherine the Great in 1764 and opened to the public in 1852.

The Hermitage has the largest collection of paintings in the world.

Sixty cats are allowed to roam the massive storerooms of the Hermitage to protect the priceless works of art from gnawing rodents.

YOU PAID HOW MUCH FOR THAT?

The most expensive painting ever sold was Jackson Pollock's four-foot-by-eight-foot board with paint dribbled on it—*No. 5, 1948*—for $140 million in 2006. William de Kooning's 1953 painting *Woman III* went for $137.5 million, also in 2006. Both paintings were sold by record executive David Geffen.

The third most expensive painting ever sold is Gustav Klimt's 1907 *Portrait of Adele Bloch-Bauer I,* which went for $135 million in 2006.

Vincent van Gogh's 1890 painting—*Portrait of Dr. Gachet*—fetched $139 million in 1990.

Pierre-August Renoir's 1876 *Bal du moulin de la Galette* sold for $131.6 million in 1990.

The most money ever paid for the work of a living artist was the $80 million David Geffen got when he sold Jasper Johns's painting *False Start* in 2006.

FROM VINCI, WITH LOVE

Leonardo da Vinci's full name is Lionardo di ser Piero da Vinci, meaning "Leonardo, son of Messer Piero from Vinci."

Since "da Vinci" means "from Vinci," it makes no sense to refer to Leonardo that way (as in the *Da Vinci Code*).

Only about fifteen of Leonardo's paintings survive today.

In 2010, secret symbols were found in the eyes of the *Mona Lisa* using magnification. There is a tiny "LV" in her right eye and a "C and E or B" in her left eye.

Leonardo was a great bird lover. He would often pay the owners of birds to release them.

Leonardo morally objected to eating meat and was a vegetarian.

He was left-handed and found it easier to write in mirror image in his notebooks.

PAINTIN' PLACE

Ultramarine blue paints used to be made with lapis lazuli from Afghanistan. They are now made from aluminum and sulfur.

The word "ultramarine" comes from the Middle Latin *ultramarinus*, which means "beyond the sea," because in the old days it was shipped to Europe over the ocean from Asia.

Yellow paints, like Indian Yellow, were made from the urine of cows that had been fed only mango leaves. These paints smelled like cow pee. Because of this, the patron who paid Vermeer to paint *Girl with the Pearl Earring*, remarked that the artist had "painted my wife with cow piss."

Vermillion red paint used to be made from mercury.

Iron filings are used to make black and red pigments.

PEACE, MAN

The peace symbol was designed by British artist Gerald Holtom for a 1958 English antinuclear protest march.

The symbol is a stylized representation of the semaphore signals for "N" and "D," short for "nuclear disarmament."

MONEY MATTERS

BUDGET BUSTERS

The top three most expensive terrestrial objects ever created are all dams—Itaipu Dam in Brazil/Paraguay cost $27 billion, Three Gorges Dam in China cost $25 billion, and the James Bay Project in Canada cost $13.8 billion.

American CVN-78 class aircraft carriers go for $8.1 billion a pop, while the Alaska Pipeline also cost $8 billion.

The ITER experimental fusion reactor under construction in France is expected to cost $6.5 billion and the CERN Large Hadron Collider on the French-Swiss border cost $6 billion.

The most expensive stadiums in the world are the New Meadowlands Stadium in New Jersey, estimated to have cost $1.6 billion, and Wembley Stadium in England, built for $1.57 billion.

ALL THAT GLITTERS

A vending machine that dispenses gold coins and small gold bars in denominations ranging from $122 to $1,400, depending on the real-time price of gold, debuted in Boca Raton, Florida, in 2010. The machine, which takes U.S. dollars, automatically updates the gold prices every ten minutes to keep current with changing gold rates.

CAN YOU SAY "PRE-NUP"?

The following are the most expensive divorces in history:

Rupert Murdoch's divorce from Anna Murdoch set him back a tidy $1.7 billion.

Adnan Khashoggi (an Arab arms dealer) wrote a check for $874 million to his wife at their parting.

Tiger Woods had to shell out $750 million to keep his scorned wife, Elin Nordegren, quiet after she dumped him.

Craig McCaw (a cellular phone tycoon) paid Wendy McCaw more than $460 million in their divorce settlement.

Michael Jordan and Neil Diamond each lost $150 million when their marriages went south.

SPOUSAL SECRETS

Eighty percent of spouses report making secret purchases. Nineteen percent have credit cards their spouse doesn't know about.

RAINY DAY

The state where the highest percentage of people have set aside an emergency fund to cover three months' worth of expenses is New Jersey. The state where the lowest percentage of folks have done so is Oklahoma.

WHO NEEDS ASPIRIN?

Research has found that the handling of cash lowers a person's perception of pain.

SEXISM SELLS

In days gone by, some advertising companies were rather sexist in their approach. For example:

A vintage Del Monte ketchup ad read, "You mean a *woman* can open it?"

An early Subaru ad touted their new GL Coupe as being, "Like a spirited woman who yearns to be tamed."

A Kellogg's Pep vitamins print ad once quipped, "So the harder a wife works, the *cuter she looks.*"

Another print ad read, "He wears the *cleanest* shirts in town . . . his 'Missus' swears by TIDE."

The Kenwood Chef kitchen mixer touted, "The Chef does everything but cook—that's what wives are for."

Tipalet cigarettes advised men to "Blow in her face and she'll follow you anywhere."

BLOWING SMOKE

Santa Claus once endorsed Pall Malls.

Camel once boasted, "More doctors smoke Camels than any other cigarette."

AMERICAN CHOICE AWARDS

The votes are in and here are America's favorite products:

The bestselling brand of beer is Bud Light.

Aquafina is the number one bottled water.

Duracell Coppertop is the leading battery brand.

Crest brushes the other toothpastes away.

Coke Classic refreshes more Americans than any other soft drink.

Lay's are the chips that the most Americans cannot eat just one of.

Bounty paper towels soak up the competition.

Breyers ice cream leaves the competition cold.

Pedigree pet food is top dog.

Meow Mix is found to be *purr*fect by a wide margin.

Chips Ahoy is the cookie of choice for dunking in a glass of milk.

Honey Nut Cheerios bowls over the competition.

Folgers coffee wakes up the most Americans.

AND STILL THE ALUMNI ASSOCIATION WANTS A DONATION

It takes the average graduate of a four-year college until the age of thirty-three to earn back the amount of money paid out in loans and lost wages while in school.

REDISTRIBUTION OF WEALTH

If all the U.S. currency in circulation were gathered together and distributed equally, each American would get $3,051.

BOEING ALL THE WAY

William Boeing made his initial fortune in the lumber business. When he started his aviation company, he had his first planes built from local spruce wood using shipbuilders from the Seattle area.

MONEY IS LIKE MANURE

Jean Paul Getty, who made his billions in the oil business, once said, "The meek shall inherit the Earth, but not the mineral rights." He also said, "Money is like manure, you have to spread it around or it smells."

Getty made twenty-one different wills, as those close to him went in and out of favor.

MAGNANIMOUS MAGNATE

Even as a teen, before he became rich, John D. Rockefeller donated 6 percent of his salary to charity. At age twenty, he upped it to 10 percent.

THE WITCH OF WALL STREET

Nineteenth-century multimillionaire Hetty Green was known as "The Witch of Wall Street."

Although Green was one of the richest women in the world, she was an incredible miser who only bought day-old bread.

Green once dressed her son in rags and tried to get him free treatment for a broken leg at a clinic for the poor. The boy's leg was eventually amputated.

She also refused to have a hernia treated because it would have cost her $150.

WAYNE'S WORLD

Wayne Newton's house has a crystal staircase, South African penguins, and a Renoir.

WARREN'S WAY

Warren Buffett is a distant cousin of singer Harry Chapin.

As a child, Buffett made money selling gum and soda door-to-door. As a teen, he took a thirty-five-dollar deduction on his taxes for the bicycle and watch that he used on his paper route.

In 2006, Buffett disowned his son Peter's adopted daughter, Nicole, after she appeared in a documentary about the growing disparity between rich and poor in America.

GOLDEN GATES

When he was in eighth grade, Bill Gates scored the highest in the state of Washington on a statewide math test.

Bill Gates's mansion, which is known as Xanadu 2.0, is sixty-six thousand square feet.

Visitors at the house wear pins that tell the home's electronics what kind of music, temperature, artwork, and lighting the guest prefers as they move from room to room.

Gates pays more than $1 million a year in property taxes.

FOURTEEN TRILLION AND COUNTING

The Bureau of Public Debt is a small agency within the U.S. Department of Treasury that borrows the money needed to operate the federal government by selling Treasury bills, notes, and bonds. They also are the ones keeping track of the national debt.

On January 1, 2011, the U.S. national debt was $14,025,215,218,708.52.

The Bureau of Public Debt accepts contributions to lower the debt on their website. They do take credit cards.

FUNNY MONEY

After the Civil War, nearly half of the money in the United States was counterfeit.

PATENTLY TRUE

IBM (International Business Machines Incorporated) was granted 4,887 U.S. patents in 2009. Samsung had 3,592, Microsoft had 2,901, Canon had 2,200, and Panasonic had 1,759.

Japan was the foreign country with the most U.S. patents in 2009, with 35,501.

EIGHT MAIDS A-MILKING

The cost in today's dollars to purchase the gifts mentioned in the classic Christmas song "The Twelve Days of Christmas" is about $23,439.

IN THE PINK

The most expensive jewel ever sold at auction was a 24.78-carat pink diamond that went for a record $46,158,674 through Sotheby's in 2010.

The world famous 45.52-carat blue Hope Diamond is believed to have been cut from a gem stolen from the Crown Jewels of French King Louis XV during the French Revolution, known as the French Blue.

The Hope Diamond passed through many hands before being donated to the Smithsonian Natural History Museum in Washington, DC, in 1958. Its owner sent the priceless stone to the museum through the U.S. mail in a plain brown paper bag.

NUMBER ONES

Nokia is the number one cell phone maker.

Honda is the leading maker of motorcycles.

PLEASE REVERSE THE CHARGES

In 1927, a phone call from New York to London cost seventy-five dollars for the first three minutes.

KEEP THE RECEIPT

Approximately 10 percent of Christmas gifts are returned to the store.

REMEMBER WHEN?

TRAGIC THEATER

Ford's Theatre closed the night President Abraham Lincoln was assassinated there. The theater was purchased by the U.S. government and an order was issued that the building should never again be used as a place of amusement. For many years, Ford's Theatre was used as a government office building. In 1893, the front of the building collapsed, killing twenty-two office workers and injuring another sixty-eight. The building sat empty from 1931 until 1968, when it was renovated and reopened as a theater.

EN GARDE

Abraham Lincoln was challenged to a duel by political rival James Shields in 1842 because of derogatory letters Lincoln had written to Illinois newspapers about him. The six-foot-five-inch-tall Lincoln chose swords as the dueling weapons, knowing his five-foot-nine-inch-tall opponent wouldn't be able to get near him. Upon seeing Lincoln's great reach advantage, Shields agreed to accept an apology instead.

PETER THE GROUCHY

Russian tsar Peter the Great stood nearly seven feet tall.

Peter was officially known as Peter I, but he renamed himself "Peter the Great."

Peter suffered from a facial tic and uncontrollable rolling of the eyes.

At the age of thirteen he organized real war games using real boys. During one such game, twenty-four of his "friends" died.

Peter demanded that all nobles shave off their beards. If he encountered a noble with a beard, he ripped it out with his bare hands. Nobles were also forbidden to marry until they learned to read.

The Russian draft started by Peter required serfs to serve twenty-five-year "enlistments."

He had his son Alexei beaten to death when he suspected him of treason.

FATHER OF THE YEAR?

Ivan the Terrible beat his pregnant daughter-in-law because he didn't like the way she was dressed, possibly causing her miscarriage. He later hit his son and heir in the head, killing him.

THE BIG SPILL

Only 8 percent of the oil spilled off the coast of Alaska in the *Exxon Valdez* disaster was recovered. Fifty percent contaminated beaches, 22 percent formed tar balls, and 20 percent evaporated.

Microbes consumed about 99.9 percent of the natural gas released in the 2010 BP gulf oil spill.

SHARK BAIT

The sinking of the USS *Indianapolis* by a Japanese submarine on July 30, 1945, resulted in the most casualties from a single incident in the history of the U.S. Navy.

Of the 1,196 men on board the *Indianapolis*, about 300 went down with the ship, which sank in just twelve minutes. Approximately 880 men ended up in the sea, most with no lifeboat or life jacket.

It was four days before anybody realized the *Indianapolis* had not returned to port. By this time, only 321 sailors remained alive, most dying from exposure and shark attacks. It is believed that the oceanic whitetip sharks that feasted on the crew were responsible for the largest shark attack in human history.

The captain of the *Indianapolis*, Charles Butler Mc-Vay III, who survived, was later court-martialed for not "zigzagging" the ship as was standard protocol in submarine-infested waters. Such was McVay's guilt that he shot himself to death with his navy service revolver on his front lawn after retiring in 1949.

TWO POPES ARE BETTER THAN ONE

In 1309, Pope Clement V, who was French, moved the papacy from Rome to Avignon, France, where it remained for seventy years. When Pope Gregory XI moved it back to Rome, in 1376, French cardinals revolted and elected their own pope, known as an antipope.

SPRY SPY

Dutch exotic dancer Mata Hari's real name was Margaretha Zelle MacLeod. She learned to perform native dances while living in Java. Her Javanese friends gave her the name "Mata Hari" meaning "the eye of the dawn."

Hari was executed by a firing squad in 1917 for being a German spy, a charge that most historians today doubt.

YOU HAVE THE RIGHT TO REMAIN SILENT . . .

Ernesto Miranda was arrested and found guilty in 1963 of being a serial kidnapper and rapist of young women in Phoenix, Arizona, but his attorney appealed the verdict on the grounds that the police had not told him of his right to an attorney when he was arrested and the confession he had made should have been inadmissible evidence. His appeal went all the way to the U.S. Supreme Court, who decided 5–4 that his rights had been violated, and since then all police are required to read those arrested their Miranda rights.

The State of Arizona retried Miranda in 1967 with new evidence—he had confessed to an ex-girl-friend—and he was sentenced to twenty to thirty years.

Miranda was paroled in 1972 and died after being knifed in a bar fight in 1976.

Ironically, the man police arrested for stabbing Miranda in the bar fight exercised his right to remain silent after being read his Miranda rights. The guy was released from custody and never seen again.

THE DIRTY THIRTIES

The dust bowl years, 1930 through 1936, were known as the "dirty thirties."

Some dust storms were so big that they blackened the skies over New York City.

One massive dust storm in the Great Plains had a width of two hundred miles.

MUMMY DEAREST

Over a period of 2,500 years, the ancient Egyptians produced roughly 500 million mummies.

One thousand yards of cloth were used to mummify a person.

King Tutankhamen ruled from the age of nine until he died at eighteen. He was a ruler of little consequence, but became famous because his tomb was found intact, which is extremely rare.

KING ME

Every English monarch since the time of William the Conqueror in 1066, except Edward V and Edward VIII, has been crowned in Westminster Abbey.

In 1936, England had three different kings—George V, Edward VIII, and George VI.

There hasn't been a king of England since 1952.

Diane de Poitiers, mistress of sixteenth-century French King Henry II, is believed to have died from ingesting too much drinkable gold—thought at the time to preserve youth.

The last Welsh person to be the Prince of Wales was one Owen Glyndŵr in 1412.

WAYWARD SON

Davy Crockett ran away from home when he was about twelve and didn't return for three years.

ISOLATION STATION

The word "quarantine" comes from the Venetian Italian *quadranta giorni*, meaning "forty days." This is the length of time that ships were required to stay in a harbor before coming ashore during the Black Death.

The *Apollo 11* astronauts were put into quarantine for two days after returning from the moon to make sure they hadn't brought back some mysterious diseases.

BODY SNATCHERS

In 1924, two youths from wealthy Chicago homes— Nathan Leopold and Richard Loeb—kidnapped and killed fourteen-year-old Bobby Franks because they wanted to commit the "perfect crime." They were caught when Leopold's eyeglasses were found near where they dumped the body. Both were sentenced to life imprisonment, plus ninety-nine years. Loeb was later killed in prison and Leopold was paroled in 1958.

In 1933, Thomas Thurmond and John Holmes abducted and killed twenty-two-year-old Brooke L. Hart in San Francisco, California. After their arrest, a mob of several thousand people broke into the county jail where they were being held and lynched the pair. The lynching was announced the day before on local radio stations and broadcast live. California governor James Rolph refused to dispatch the National Guard to protect Thurmond and Holmes and publicly praised the lynching.

Coors beer heir Adolph Coors III was kidnapped and murdered in Colorado in 1960.

Julio Iglesias's father was abducted in 1985, but was later found alive and well.

In 2005, police thwarted a conspiracy by a housepainter to kidnap David Letterman's son, Harry Joseph Letterman.

In 1971, Yoko Ono's second husband, Anthony Cox, abducted their daughter. Ono didn't see the girl again until 1998.

In 1997, a ten-day-old infant—Delimar Vera—was presumed incinerated in a fire that destroyed her Philadelphia house. In 2004, the girl's mother, Luz Cuevas, was struck by the resemblance of a six-year-old girl at a family friend's birthday party to her other children. Cuevas took some strands of hair from the child and DNA testing proved her to be Delimar. The kidnapper, who had set the fire to cover up the abduction, eluded police until 2009.

ANARCHISTS AND MADMEN

Charles Guiteau, the man who mortally wounded President James A. Garfield in 1881, had written a speech supporting Garfield and believed he was responsible for his election. Afterward, Guiteau demanded an ambassadorship and became enraged at Garfield when one wasn't offered. As revenge, he shot Garfield twice in the back at a railway station in Baltimore.

The assassination of Garfield was well thought out. Guiteau went so far as to check out the accommodations of the jail that he would likely spend time in to determine its suitability.

Guiteau was hanged nine months after the assassination and part of his brain is now on display at the Mütter Museum in Philadelphia.

Teddy Roosevelt was shot and wounded by a mentally ill man just before giving a campaign speech in Milwaukee in 1912. The bullet was slowed down by a metal eyeglass case and a folded fifty-page speech that he had in his jacket pocket. Although the bullet had lodged in his rib cage, Roosevelt went ahead and gave the ninety-minute speech before seeking medical attention. The bullet was never removed.

President-elect Franklin D. Roosevelt was shot at by anarchist Joseph Zangara in Miami in 1933. FDR was saved when a woman grabbed the gunman's arm, causing a bullet to fatally wound Chicago Mayor Anton J. Cermak. Four others were also wounded in the incident.

Senator Huey P. Long was gunned down and killed by Dr. Carl Austin Weiss, the son-in-law of a political opponent, in 1935 inside the Louisiana State Capital.

Alabama governor George Wallace was shot by Arthur Bremer while campaigning in Laurel, Maryland, in the Democratic presidential primaries in 1972. Wallace was hit five times and spent the rest of his life wheelchair-bound. Bremer was motivated by fame, not politics. He was paroled from prison in 2007. The Wallace assassination attempt inspired the screenplay for the movie *Taxi Driver*.

REMAIN RESOLUTE

The HMS *Resolute* was a British ship that became trapped in the ice in the Canadian arctic and abandoned in 1854. An American whaling ship found her and towed the *Resolute* back to Groton, Connecticut, where the U.S. Congress paid for the ship to be repaired and returned to Britain as a show of goodwill. When the British decommissioned the vessel in 1879, four desks were made from her timbers, one of which was presented to President Rutherford Hayes in 1880. Known as the Resolute desk, it is used by Barack Obama in the Oval Office and was featured in the plotline of the movie *National Treasure 2: Book of Secrets*.

SUPERSIZE IT

The first practical fax machine weighed forty-six pounds.

The first automatic answering machine stood three feet tall.

The first bar code scanner was the size of an office desk.

The first hearing aid was the size of a suitcase.

The first electronic calculator was the size of a small room and the first "compact" calculator was the size of a typewriter.

TAKING A TOLL

Toll roads date back to Babylon in 700 BC.

PILGRIM POWER

The *Mayflower* took sixty-six days to sail from Plymouth, England, to Plymouth, Massachusetts.

Two babies were born en route.

The ship's destination was to have been the mouth of the Hudson River, near present-day Manhattan.

The *Mayflower* landed on November 21, 1620, just as winter was setting in. The Pilgrims spent the winter on the ship and moved ashore in March.

Within the first three months after landing, half of the Pilgrims were dead.

The *Mayflower* passengers included Pilgrims and merchants. At first, the colony operated under a communal agricultural system. When this failed, each person was granted their own parcel of land.

Pilgrims and Puritans were both Protestants, but while the Pilgrims left the church, Puritans wanted to purify it.

Puritan families averaged about eight children.

The first Thanksgiving lasted three days.

One in ten Americans today can trace ancestry to someone on the *Mayflower*.

SLAVING AWAY

The first Africans were brought to the British Colonies of North America by the Dutch in 1619. These first twenty landed at Jamestown, Virginia, and were treated as servants, not slaves. They were able to gain their freedom and own land.

INSPECTION STATION

The immigration depot at Ellis Island was for third-class passengers. First- and second-class passengers were processed by officials while still onboard their ships.

It took between three and five hours to be processed through Ellis Island.

About 98 percent of arriving immigrants were allowed into the country.

Ellis Island processed about five thousand people a day.

The busiest year for Ellis Island was 1907, with 1,004,756 immigrants processed.

STATUESQUE LADY

It took two hundred men, working seven days a week, nine years to build the Statue of Liberty.

The Statue of Liberty has broken chains under her feet that represent breaking the chains of tyranny. They can only be seen when viewed from above.

HOMELAND INSECURITY

Up until September 11, 2001, the deadliest terrorist attack in the United States was the bombing of the J.P. Morgan Bank Building in New York, in September 1920. It killed thirty-nine and injured three hundred. The perpetrator was never identified.

On March 1, 1971, the Senate wing of the U.S. Capitol Building was bombed by the radical group Weather Underground. In 1975, the Weather Underground bombed the U.S. State Department building in Washington, DC. There were no injuries in either attack.

In 1975, radicals from the Puerto Rican FALN bombed Fraunces Tavern in New York City, killing four and wounding fifty-three.

The truck bomb that was detonated by al-Qaeda in the parking garage beneath the World Trade Center in 1993 killed six people.

🌰 PICK UP THE PIECES

The largest crime scene ever was the eight-hundred-square-mile debris field of Pan Am Flight 103, which was destroyed by a bomb over Lockerbie, Scotland, in 1988.

More than ten thousand pieces of evidence were collected in the countryside.

Forensic scientists were able to determine the exact suitcase that the explosives were placed in.

A man named Jaswant Basuta just missed the flight after checking in his bags and lingering too long in the airport bar. The music group the Four Tops were supposed to be on the plane, but fortunately overslept. Johnny Rotten of the Sex Pistols and his wife, likewise, missed the flight because they were delayed.

STALIN'S STANDOFF

During World War II the Germans laid siege to Leningrad for 872 days, resulting in the death of up to 1.5 million Soviets.

CHANGING TEAMS

Japan and Italy were on the side of the Allies, fighting *against* Germany in World War I.

Along with Germany, Italy, and Japan, Albania, Bulgaria, Finland, Hungary, Romania, and Thailand were part of the Axis powers in World War II.

BETTER LATE THAN NEVER

Mongolia was the last country to join the Allies during World War II, on August 9, 1945, just a few days before the war ended.

TREATY ON
THE ORIENT EXPRESS

The original Orient Express ran from Paris to Istanbul.

The Germans signed the document surrendering to the Allies at the end of World War I in a train car on the Orient Express. The French kept this car on display in Paris until the Nazis defeated France in 1940. Hitler had the car hauled to the exact spot where the Germans were forced to surrender twenty-two years earlier. There, he gleefully accepted the French surrender. Four years later, when it looked like Germany was going to lose another war, Hitler had the car destroyed to avoid a repeat performance.

Today's Orient Express is a luxury train that goes through twenty-two thousand bottles of champagne a year.

ON THE FRONT

The Crimean War (1853–56), between Russia and allied European forces, was the first to be covered by newspaper reporters in the field.

DOPEY DESPOTS

Saddam Hussein fancied Cheetos Puffs and requested them regularly while being held prisoner before his execution.

Hitler was a vegetarian. His personal physician gave him daily injections containing animal by-products, such as placenta, extract of seminal vesicle and prostate, liver, pancreas, and adrenal gland.

Idi Amin had a *Tom and Jerry* cartoon collection.

Khmer Rouge leader Pol Pot had the entire population of Cambodia's cities resettled to become slave laborers on collective farms. As a result, between 1976 and 1979, approximately 2 million people died, about 20 percent of the country's population.

Pol Pot had intellectuals systematically murdered for being "useless eaters." Merely wearing glasses was enough to qualify one as an intellectual.

STANDING FAST

The only Southern lawmaker not to leave the Union when the Civil War broke out was U.S. Senator Andrew Johnson of Tennessee. As vice president, he became president of the United States after Abraham Lincoln was assassinated.

NORTHERN EXPOSURE

The northernmost attack on the United States by Confederates during the Civil War was at St. Albans, Vermont, in 1864. Ben Young led twenty-one escaped rebel POW cavalrymen who had fled to Canada on a raid of the town's three banks, fifteen miles south of the border. They stole $208,000 and tried to burn down the town, but heavy rains thwarted their attempt. The raiders returned to Canada with the money, where they were promptly arrested and forced to give back what was left of the loot.

POLITICS AS USUAL

To avoid the bad PR, Abraham Lincoln had General Sherman hold off on his March to the Sea across Georgia, until he was reelected president in 1864.

BIG BANGS

On April 16, 1947, a cargo ship containing ammonium nitrate exploded in the Port of Texas City, Texas. The detonation resulted in a chain reaction of fires and explosions involving adjacent ships, warehouses, and chemical and oil tanks. Two thousand buildings were leveled and at least 581 people killed. The blast shattered windows in Houston, some forty miles away, and threw the ship's two-ton anchor over one and a half miles away.

> In 1917, the French ammunition ship SS *Mont-Blanc* collided with another ship in the harbor at Halifax, Nova Scotia, caught fire, and exploded. The blast leveled buildings for two square kilometers and killed about two thousand people. With a force of 2.9 kilotons of TNT, this blast may have been the most powerful accidental explosion ever.

During the Cold War, the Soviet Union tested a nuclear bomb that was three thousand times more powerful than the bomb dropped on Hiroshima.

More than one thousand nuclear tests were conducted in the state of Nevada during the Cold War.

LUCKY LINDY

Charles Lindbergh bailed out of crashing airplanes four different times.

Lindbergh and friend Henry Ford were both anti-Semites.

LUCKY LADY

One Violet Jessop holds the distinction of having survived the sinking of the RMS *Titanic* in 1912, the sinking of *Titanic*'s sister ship, the HMS *Britannic* in 1915, and the collision of the third sister ship, the RMS *Olympic*, with a navy vessel in 1911.

SMOKE ON THE WATER . . .

Ohio's Cuyahoga River has caught fire at least thirteen times since 1868. The 1969 fire provided the impetus for the U.S. Congress to pass the Clean Water Act.

. . . FIRE IN THE SKY

In 1988, one-third of Yellowstone National Park was consumed by flames. In one day, 150,000 acres burned.

MORMON MASSACRE

On September 11, 1857, in what is known as the Mountain Meadows Massacre, Mormon militiamen ambushed a wagon train of families from Arkansas passing through the Utah Territory. After persuading the travelers to surrender, the Mormons killed all the men, women, and children over the age of eight to eliminate any witnesses. The children under eight were distributed to Mormon families. In all, 120 people were executed.

BAD GIRLS

Twelve women have been executed in the United States since capital punishment was reinstituted in 1976.

TALK ABOUT HOLDING A GRUDGE

In 2010, Pope Benedict XVI became the first pope to make an official "state" visit to Britain since King Henry VIII broke away from the Catholic Church five hundred years ago.

FELONIOUS FAMILY

In 1969, Daniel Ellsberg stole papers from the Pentagon that showed Presidents Truman, Eisenhower, Kennedy, and Johnson lied about escalating the war in Vietnam.

Ellsberg had his thirteen-year-old son and ten-year-old daughter photocopy the seven thousand pages of documents he then sent out to newspapers and members of Congress.

BROKEN ARROWS

"Broken Arrow" is U.S. military parlance for an acciden-tal event involving nuclear weapons.

In 1956, an air force Stratojet carrying nuclear weap-ons disappeared over the Mediterranean and was never found.

In 1958, a thermonuclear bomb was dumped a few miles off the coast of Georgia by an American B-47 bomber plane that collided with a fighter jet during war games. The H-bomb has never been recovered.

In 1965, a Navy A-4 Skyhawk airplane fell off the el-evator of the USS *Ticonderoga*. Its pilot and nuclear bomb sank to the bottom of the Pacific Ocean near Okinawa.

FIRST FUMBLES

The "football" is the briefcase that contains the instruc-tions to launch a nuclear attack that is always with the U.S. president.

In 2000, Bill Clinton lost the nuclear launch card that has the code numbers on it that allow the commander in chief to access the "football." It was months before the Pentagon realized that the card had gone missing and issued Clinton a new one.

Jimmy Carter apparently was just as careless. He once sent a suit to the cleaners with the card in a pocket.

THEY DON'T BUILD THEM LIKE THAT ANYMORE

The Colosseum in Rome was built almost two thousand years ago and much of it still remains standing. Giants Stadium in New Jersey, by comparison, only lasted thirty-four years before it was torn down.

WITH FRIENDS LIKE THESE . . .

Even though England and France were both Allies in World War II, on July 3, 1940, the British Navy destroyed much of the French fleet in the port of Mersel-Kebir in French Algeria so that it would not fall into German hands. The French lost 1,297 sailors during the battle.

In 1967, during the Six-Day War, Israeli jets and torpedo boats attacked the USS *Liberty* in international waters near the Sinai Peninsula, killing thirty-four American sailors and wounding 170.

NOT KNOWING WHEN TO QUIT

Germany did not surrender for a week after Adolf Hitler killed himself at the end of World War II.

Hitler's dog at the time of his death was a German shepherd named Blondi. Hitler had the dog killed to test the suicide pills he had been given by a doctor to use for his own demise.

LETHAL WEAPON

The AK-47 is the most prolific weapon ever produced. About 140 million of these automatic rifles have been made to date.

The AK-47 has killed more people than any other weapon in history.

MURDER INC.

In 1984, a pesticide plant operated by Union Carbide accidentally spewed forty tons of deadly methyl isocyanate gas into the air in Bhopal, India, ultimately killing fifteen thousand people.

IN THE PINK

Up until the 1940s, pink was typically considered the color for boys and blue the color for girls.

GREAT MOMENTS IN HISTORY

A man invented the modern tampon in 1933.

The steering wheel was invented in 1894. Before this, early autos were steered with a joystick-type handle.

In 1937, the Chelsea Baby Club in London distributed metal cages that were hung outside of tenement windows to give babies fresh air and sunshine.

Early monks in silent orders invented the first sign language.

When the Huns attacked what is present-day Italy in the mid-400s, many people from the countryside sought refuge on the islands in the Lagoon of Venice. This settlement later became Venice.

The oldest ski ever found was one dating to 2500 BC that was unearthed in Scandinavia.

The only U.S. Army survivor at Custer's Last Stand was a horse named Comanche, who was later stuffed and is now on display at the University of Kansas.

On the Roman holiday of Saturnalia slaves and masters switched roles.

The Romans didn't use chariots for warfare, but for racing. Chariots also were used in ceremonial processions, sometimes pulled by dogs, ostriches, or even tigers.

ACKNOWLEDGMENTS

As always, I must thank my editor and the creator of this wonderful series of Useless Information books—Jeanette Shaw. Without her efforts, you wouldn't be reading this book. The copyeditor, Emma Hinkle, who had the monumental task of verifying the accuracy of a book packed with so much information, has my gratitude, as does Sarah Romeo, who put together the most humorous cover so far in the series. Last but not least, many thanks to my "super" literary agent, Janet Rosen, who handles the business end of my career so I can concentrate on writing.

The
#1 *NEW YORK TIMES*
Bestselling Series

COMING JUNE 2012
A special edition for
Young Readers!